American Dimestore

Toy Soldiers and Figures

Don Pielin, Norman Joplin and Verne Johnson

4880 Lower Valley Road, Atglen, PA 19310 USA

Library of Congress Cataloging-in-Publication Data

Pielin, Don.
American dimestore toy soldiers and figures / Don Pielin, Norman Joplin, and Verne Johnson.
p. cm.
ISBN 0-7643-1189-1 (hardcover)
1. Military miniatures--Collectors and collecting--United States--Catalogs. 2. Action figures (Toys)--Collectors and collecting--United States--Catalogs. I. Joplin, Norman. II. Johnson, Verne. III. Title.
NK8475.M5 P485 2000
688.7'2--dc21
00-009479

Copyright © 2000 by Norman Joplin

All rights reserved. No part of this work may be reproduced or used in any form or by any means—graphic, electronic, or mechanical, including photocopying or information storage and retrieval systems—without written permission from the copyright holder.
"Schiffer," "Schiffer Publishing Ltd. & Design," and the "Design of pen and ink well" are registered trademarks of Schiffer Publishing Ltd.

Designed by Bonnie M. Hensley
Type set in BankGothic Md BT/Korinna BT
ISBN: 0-7643-1189-1
Printed in China
1 2 3 4

Published by Schiffer Publishing Ltd.
4880 Lower Valley Road
Atglen, PA 19310
Phone: (610) 593-1777; Fax: (610) 593-2002
E-mail: Schifferbk@aol.com
Please visit our web site catalog at **www.schifferbooks.com**

In Europe, Schiffer books are distributed by Bushwood Books
6 Marksbury Avenue Kew Gardens
Surrey TW9 4JF England
Phone: 44 (0) 20-8392-8585; Fax: 44 (0) 20-8392-9876
E-mail: Bushwd@aol.com
Free postage in the UK. Europe: air mail at cost.

This book may be purchased from the publisher.
Include $3.95 for shipping. Please try your bookstore first.
We are always looking for people to write books on new and related subjects.
If you have an idea for a book please contact us at the above left address.
You may write for a free catalog.

Contents

Acknowledgments 4

Introduction 5
    History 5
    Company Histories 6
    List of Manufacturers Mentioned in the Text 8
    Terminology 9
    Toy Soldier and Figure Sizes 10
    Types of Materials 10
    Valuations Guide 10

Part I: American Social History in Toys

1. Early American Manufacturers 14
2. The American Family 20
    African American Family 22
    Amish American Family 23
    American Family Wedding 24
3. Railroad 26
4. Farm 33
5. Indians 44
6. Cowboys 53
7. Uniforms 65
8. American Legion 70
9. Holidays 71
    Christmas 71
    Winter 74
    Halloween 76

Part II: Military

10. Camp 80
11. Medical 85
    Nurses 85
    Doctors and Wounded 86
    Stretcher Parties 89
12. Officers 92
13. Parade Figures 95
14. Musicians 104
15. Flagbearers 109
16. Signal Flags 112
17. Calvary 114
18. Communication 118
19. Firing Figures 120
20. Machine Gunners 126
21. Grenade Throwers 132
22. Advancing and Charging 134
23. Flame Throwers and Bazookas 140
24. Gas Mask Figures 142
25. Anti-aircraft 144
26. Anti-tank 148
27. Searchlights 153
28. Motorcycles 155
29. Accessories 157
30. Ski Troops 159
31. Military School Cadets 160
32. Air Force 162
33. Parachutists 164
34. Navy 166
    Pirates 169
    Divers 170
35. Marines 171
36. Foreign Troops 173
37. Miscellaneous 178

Part III: Historical and Non-Military

38. Historical Figures 182
39. Circus 193
40. Disney 195
41. Character Figures 197
42. Space and Science Fiction 201
43. Souvenir and Novelty 204
44. Sports 210
45. Army Vehicles 218
46. Aircraft 226
47. Ships 228
48. Box Art Work, Catalogs, Advertising, Molds 230

Subject Index 240

Acknowledgments

The majority of figures photographed for the book came from the collection of Don Pielin, Co-Authors Verne Johnson and Norman Joplin also provided figures from their specialist collections. To achieve the comprehensive coverage that was required, many fellow collectors were most supportive and supplied figures from their collections.

Special thanks are therefore due to: Debbie Bednarek, Marv Breitlow, Ron Cadeiux, Deb and Ron Eccles, Don Hovde, Roger Johnson, Sheila Joplin, Joe Kunzelmann, Dick Pielin and Kent Wood.

Special thanks also to Kent Wood for the photography, Carol Runkle for typing the manuscript, and Doug Congdon–Martin and the staff at Schiffer Publishing.

Introduction

American Dimestore toy soldiers and figures is the first full color reference book devoted to toy soldiers and figures that were available in the U.S. "Five-and-Dime" stores. Divided into thematic sections rather than by manufacturer, it enables both experienced and new collectors to quickly locate and identify figures.

A reference book of this nature can never be fully comprehensive in content and it is inevitable that some items have been omitted because of their rarity and could not be located for photography.

History

Toy soldiers first appeared on America's retail market towards the end of the 19th and the beginning of the 20th century. They were mainly imports from Europe where the toy soldier industry was well established. France had led the way. Lucotte, which dated back to 1760, produced what is believed to be the first commercial toy soldiers, Mignot, still in existence today, took over the Lucotte company. Mignot was joined by a number of German manufacturers in the 19th century, the best known being Heyde. William Britain revolutionized toy soldier manufacturing in England when, in 1893, the hollow cast method of production was perfected. This not only eliminated British imports from France and Germany, but also proved to be the start of trade between Great Britain and the U.S.A. When William Britain's company started to supply the American soldier company with hollow cast figures for the Beiser Military Shooting Games, a number of enterprising American-based manufacturers started producing their own similar lines. Influenced in design by the European products on the day, the scale of these figures was a constant 54 mm or 2-1/4 inches. Up until 1893, all toy soldiers were made of solid lead. American-made toy soldiers up to this point had not yet achieved their own identity.

The popularity of the electric toy train during the 1920s had a bearing on this unique identity being achieved. Standard gauge scale toy trains produced by Lionel, American Flyer and Ives, along with European manufactured toy train imports, brought about the advent of the 3-inch toy figure. The 2-1/4-inch European style scale became redundant as it proved too small to compliment the size of American toy trains. Eventually, the 3-inch toy soldiers and figures would create their own identity and capture a large share of the American toy market which in some ways surpassed the popularity of the toy train. German manufacturers probably influenced the American manufacturers during the 1920s with their 7 cm figures made of composition material. It was not long before a number of American toy soldier companies began producing their own hollow or slush cast ranges of lead figures, in a way amalgamating all of the ideas presented to them by their European counterparts.

It is probably fair to assume that Edward Jones of Chicago first had a vision indicating that the size of figures should be increased from 2-1/4 inches to 3 inches. Jones, who had his molds manufactured in England and was in awe of the British manufacturers' hollow cast methods, could therefore be called the pioneer of American made 3-inch (dimestore) toy soldiers and figures.

Barclay and Manoil were the most prolific manufacturers in lead, while Grey Iron's cast iron products held a strong place in the market. Like any industry success breeds success and it was not long before a number of other companies started to produce their own unique lines.

The "Five-and-Dime" stores (dimestores) were the perfect market place for these toy soldiers and figures. Priced accordingly and sold along side the trains that they complemented, by the mid-1930s, most of the major "Five-and-Dime" stores would carry large ranges of toy figures.

Three-inch railroad figures were obviously influenced by and compatible with the toy trains of the day. However, manufacturers were not short of ideas when it came to deciding upon and producing new subjects. American social history itself would prove to be the inspiration for the toy soldier companies product lines produced during the next thirty to forty years. From the American family to sports figures, from cowboys and Indians to soldiers of World War One and World War Two, and from farm, historical and Christmas holiday figures to vehicles, ships and aircraft, all manner of toys were produced reflecting American social history of 1930s and 1940s America through the eyes of a child's toy.

Although the 3-inch figure became what is described in the text as standard issue figures, other scales were also produced (see list of toy soldier sizes). These included figures which complemented the 0-27 scale toy trains. Most figures were made of lead, however, composition materials and rubber were used, sometimes because of the lead shortage during World War Two. Miller, Sliktoy, Molded Miniatures and Playwood Plastics were of composition material, while rubber was used successfully by Auburn Rubber.

After World War Two, the lead toy soldier and figure market started to decline and struggled to maintain its popularity. The post-war occupation of Japan saw many Japanese products on display in the dimestores during the 1950s. The advance of technology and the advent of plastic ensured that lead figures would gradually fade from the dimestore's toy shelves. Dimestore toy soldier collecting is, along with collecting toy soldiers in general, a growth hobby in the U.S. and Europe and, in the main, an adult one creating an excuse for adults to recapture their childhood toys.

Many specialist toy soldier shows and dealers offer wide selections of toy soldiers and figures through the second hand market, and the figures have now achieved antique and collectible status.

Company Histories

All Nu Products (Faben Products post-war) was founded in the late 1930s in New York City. Their military, western and souvenir figures were beautifully designed, but had limited exposure on the market, hence the rarity of these toys.

American Alloy (also known as Toy Creations) was formed in early 1941. They produced only copies of the Tommy Toy figures but eliminated the markings under the base.

American Metal Toys, one of the few toy soldier companies founded away from the east coast, was formed in 1937 in the Chicago area. It was the first company to provide an "enemy" army with gray painted versions of the American soldiers it cast.

Arcade, best known for its cast iron vehicles, it produced limited numbers of nickel-plated soldiers as well as a few pieces of army equipment. The company only made 5 different poses and these were plated and some were sprayed with an orange lacquer to represent the "enemy."

Auburn Rubber (Aub-Rub or ARCOR) had its beginnings in the early twentieth century, but did not produce toy figures until the late 1930s. The company's line included soldiers, farm, and circus, but the animation was limited by the fact that vulcanized rubber was used in the manufacturing process. The instability of this material has led the company's production to lose favor amongst some collectors.

Authenticast. Authenticast's history can be traced from its beginning in the late 1940s to the mid-1960s when it ended operations. Both Holger Eriksson and Frank Rogers sculpted figures for the company, which produced mainly boxed sets of soldiers. Sports and farm figures were the exception.

Barclay, by far the best known and most prolific of the toy soldier makers, was incorporated in New Jersey sometime in the early 1920s. First producing 2-1/4-inch figures and later the standard 3-inch scale, the company introduced the separate tin helmet applied to their prewar standard size figures. Barclay produced a full range of soldiers, western, farm, winter and railway figures almost to the end of metal production in 1971.

Beton. Beton's early figures were cast in lead from molds made by Metal Cast. The fact that they were in business by 1935 and had switched to plastic by 1938 indicates the reason why so few of their metal figures exist.

Breslin, one of the two main Canadian toy soldier makers. The company's main line was a series of Manoil, Barclay and American Metal Toy copies. Mainly produced during and right after WWII, these figures are identifiable by rough castings and crude paint jobs as well as the word "Canada" cast on the toys.

Cassidy (Jones). Cassidy bought the molds or at least the rights to the castings from J. Edward Jones in the 1950s. Jones had advertised these 0-27 gauge train figures under the Moulded Miniatures name but probably never sold them himself.

Cosmo, another short lived New York toy soldier company. Advertising figures available from 1930, its crudely cast and painted production only lasted a few years.

Grey Iron Castings Company has the longest history of any of the toy soldier makers, beginning in 1881. The company first produced the 2-inch Greyklip Armies in 1917, and later produced standard size figures in the late 1930s. The figures have been produced off and on in Pennsylvania up through the 1980s when the company offered unpainted castings of most of their figures direct from the foundry.

HB Toys, a New York company founded by the Helm Brothers after WWII. The product was entirely made up of crude copies of Barclay figures, some modified Barclay design and the Robe man and Robe boy which were unauthorized attempts at capitalizing on the fame of Captain Marvel.

Historical Miniatures, based in New York City, made beautifully designed standard size figures of mainly historical personages. These were solid cast and had paper labels under the bases. Their production run seems to have been from just pre-WWII until the mid-1950s.

Jones (Metal Art, Moulded Miniatures). Jones had been involved in the toy soldier business from the mid-1920s. His Chicago-based Metal Art company produced both 2-1/4 and 3-inch figures of historical subjects. Later, Moulded Miniatures kept the smaller figures that had been introduced, and brought in more modern WWII figures. New subjects were added right up to the company's end in the mid-1950s.

Lincoln Log. Lincoln Logs were the creation of John Wright, son of the famous architect, who noted that Japanese children played extensively with building sets. He bought Noveltoy to produce figures to go with his log sets. During the transition many former Noveltoy figures were produced by Lincoln Log but were unmarked. Once Lincoln Log was able to begin production on their own, all figures were marked "Lincoln Log USA."

London Toy, the only other Canadian toy soldier manufacturer of any significance. They produced a rather static series of Canadian soldiers, airplanes, and vehicles. Their production was marked "London Toy Canada" and limited to the 1940s.

Manoil. Manoil Company began life in the mid 1920s, but it wasn't until the mid-1930s that they produced toys. Most of their line was marked "Manoil" and are generally identified by the robust bodies and animated attitudes. The company produced toy figures and vehicles until the late 1950s when it ceased business.

Metal Cast. Metal Cast Products Company's chief business was the making of casting molds. They sold the molds and later when they added slush casting molds to their line; they also sold castings to their customers. Their soldiers and other novelty items were unmarked, but easily identified through the many original catalogs still in existence.

Minikin, the best of the Japanese toy soldier manufacturers. Their well-designed toy soldiers were produced from the "occupied Japan" era into the early 1960s.

Miller. Miller began business in Chicago by making nativity figures and selling them as individual pieces, a practice not allowed by foreign manufacturers of the time. Their 5-inch soldiers appeared in 1950 and the company continued to turn them out along with holiday items until the end of the decade. Miller items are identified by the remains of the cardboard tab in the under-base by which they were pulled from the mold.

Molded Products. This company responded to war time shortages by producing figures made of wood flour (fine sawdust) and glue. The inconsistency in size and shape of the same figures is attributed to the variance in curing time. Beginning in New York around 1941, the company lasted until 1946 when post-war plastics took over their market. They are described in the text as composition.

Multi-Products. Multi-Products began production in the late 1930s with comic characters mainly from King Syndications and Disney. They expanded to souvenir and novelty items during the metal shortages of the War. The post-war comic craze kept them going until cheap plastics and competition from Japan put them out of business.

Noveltoy. Noveltoy began business in Chicago in 1930 and was acquired by Lincoln Log in 1933 to produce toy figures to compliment the building sets. The original designs and molds were by Henry Kasselowski (Henri Castle) who worked for Lincoln and Jones also. Early figures are marked "Noveltoy," the same figures were produced unmarked during the transition to Lincoln Log.

Paul Paragine. Paragine's toy soldiers were simple copies of a few of Manoil's figures. This small New Jersey company apparently copied Manoil but used a two-piece mold and not the more expensive three-piece mold which meant the under base was hollow. It is interesting to note that the kneeling firing figure is a copy of the folding rifle Manoil soldier.

Playwood Plastics (Transogram) took advantage of the war time metal shortage to produce a series wood composition toy soldiers. The fact that most of the two soldiers came in versions indicates either an expansion because of demand or an attempt at better designs. The New York-based branch of Transogram/Gold Medal ceased production of the figures shortly after the war.

Sheila Inc., a Cleveland-based novelty company, produced several sports figures of note. Making their figures in the late 1930s, they seem to have used both slush casting and die casting methods. The early slush cast figures originally had paper labels; the die cast figures were unmarked.

Slik-Toy. A trade name used by Lansing Company of Iowa for their toy production. Originally a metal farm toy maker they branched into plaster (gypsum) soldiers in the early 1950s. They are slightly smaller than their Miller counterparts, but just as detailed. Production of these figures ended by 1957.

Tommy Toy. Like most of the other small toy soldier companies had its start late in the 1930s in New York City. Their figures are cleanly sculpted and well-animated, coming from the hands of one of Barclay's sculptors. Tommy Toys were eventually sold to American Alloy and produced with no markings under the base.

Trico, the best known of the Japanese composition imports. They usually worked from their own designs, although several of these relate directly to European composition. The company actually refers to the importing and labeling company, not to the Japanese company who actually manufactured them.

Tru-craft. Tru-Craft began business by making 0-27 gauge train figures, later adding soldiers in the same scale. Their Los Angeles location made them one of the first West Coast manufacturers. After WWII, they made and sold 2-1/4-inch military and football figures, the company was sold to the Jack Scruby family who still produce a limit range of them.

Wilton. Wilton Company has always had its roots in east central Pennsylvania. Their production first of cast iron industrial goods and later toys and novelties, was prolific. Still in business, they and their competitor, Grey Iron (John Wright) seem to have copied each other's pieces, making late production items difficult to identify.

Japanese. Japanese makers seem to have entered the market during the period just before WWII broke out in Europe. They worked in various materials, including composition, bisque, plastic and lead alloy. By far, most of their figures were copies of already existing figures from the U.S. and Europe.

Souvenirs. Souvenir pieces were usually created for the ubiquitous market of people on vacation. Impulse buying supported these lines so that the quality of casting and authenticity of design had no bearing on sales. Retail outlets usually added the name of their location to each figure, sometimes resulting in the same items being offered as souvenirs in different locations. Most major tourist attractions and historical landmarks in North America offered souvenirs in the form of figurines. The variety, therefore, is immense and for the collector of such items the task of collecting them is never ending. No one company can be attributed to those items and although many were made in the USA, Japanese imports almost certainly were equal in production.

List of Manufacturers Mentioned in the Text

All Nu
Allied Manufacturing
American Alloy
American Metal Toys
American Soldier Company
Arcade
Auburn Rubber
Authenticast
Barclay
Barr Rubber
Beiser – *See American Soldier Company*
Best Maid
Beton
Breslin
Cassidy/Jones
Chein, K.
Christies Metal and Toy Soldier Company
Cosmo
Degay
Eureka – *See American Soldier Company*
Faben – *See All Nu*
Feix, William
Grey Iron
Gutmann, Ferdinand
H.B. Toys
Hahn, Theodore
Hill, John (Johillco), England
Home Foundry
Hubley
Jaymar
Jones
Judy Toy
Junior Caster
Kresge
Lincoln Log
London Toy
Manoil
Metal Cast
Minikin
Miller
Molded Products
Multi Products
Nifty
Noveltoy

Paul Paragine
Playwood Plastics
Ralstoy
Sail Me Company
Schneider Trench Company
Sheila
Slikite
Slik Toy
Soljertoy
Sonsco
Sun Rubber
Timpo, England
Tommy Toy
Tootsietoy
Trico
Tru Craft
Velasco Toy Company
Victory
Warren
Well Made Doll Company
Wilton

Terminology

Certain descriptions used are unique to toy soldier collecting and although not used by the manufacturers of toy soldiers, they convey the message via descriptive slang terminology. Much of the following terminology is used universally by toy soldier collectors and dealers describing toy figures. Some items may appear in or appear to belong to more than one category, i.e., Barclay Boy Scout with signal flags appears in both the signal flag and uniform sections.

Barclay toy soldiers and figures. Pod Foot, Midi Series, Mini Series.

Casting variations are noted as long stride, short stride, cast helmet, pot helmet, hollow base or flat base.

Dimestore Toy Soldiers. A term credited to Don Pielin who coined the name while compiling his first privately published book on toy soldiers. It is an all-embracing name which describes any type of toy soldier or figure available for sale at the many "Five and Dimes" (dimestores) such as W.T. Grant, Kresge, Newberry and Woolworths.

500 Series – Although a smaller size, their design related directly to the early post war 45/46 series. *See Manoil.*

45/46 – These figures were issued immediately after the war, the numbers which appear on the figures probably relate to the years in which they were designed. *See Manoil.*

Greyklip Armies – These are nickel plated cast iron figures sold in sets originally mounted on cards.

Home cast – Home cast refers to figures produced from molds sold to hobbyists during the 1930-1950 period. The molds were sold via mail order by a number of companies including Allied Castings, Home Foundry, and Junior Caster. The molds were filled with melted lead and the resultant figures were painted for personal play. Their lack of commercial recognition leaves them almost valueless in today's market. The molds themselves have, however, become quite collectible.

Long stride – Figures produced later but still pre-war, identified by more animation and greater space between the feet on the base. These terms are used to distinguish between figures in the same position but produced at different times.

Manoil toy soldiers and figures. 45/46 Series, 500 Series Skinny Series.

Midi – These are a smaller version of the pod foot soldiers, although all the designs were new, they strongly resembled the pod foot poses. Two cowboys and two Indians were also produced in this scale.

Mini – The mini name is given to the 0-27 scale railroad figures produced after the war. These figures were only made in civilian designs.

Pod Foot – Later version figure without a solid base, standing on a round "pod" under each foot are indicated by this term. They date from the mid fifties until the end of military production. The name applies to all figures in this size from the same period.

Pot Helmet – The larger "pot" style helmet which was standard issue U.S. Army helmet by 1943, indicates the 3 inch figures produced immediately after the war by redesigning the early poses with a more modern helmet. These are usually green and sometimes silver, but white was used for the musician pieces.

Pre-dimestore is a term used to describe items made either in the USA or for the American market, and are believed to have been available before the advent of five and ten cent stores.

Short stride – Early pre-war, static figures with their feet close together on the base, also includes figures from the same period even though they are not in a standing position.

Skinny Series – These were issued shortly after the 45/46 series and while the poses are sometimes similar, their design was much thinner in body, hence the nickname. These were also issued unpainted for the hobbyist to paint themselves.

Slope arms is used to describe all figures with rifles on shoulders.

Standard issue. All three inch figures, in the absence of other wording.

Uncle Sam's Defenders – Originally some of the Greyklip army painted khaki, later they were redesigned to have a more modern uniform. Those found with painted faces were early issues.

Variant is used when the same figure is finished in different paint colors.

Version is used to describe casting differences in the same basic item.

Toy Soldier and Figure Sizes

The caption descriptions correspond with the sizes listed. Some manufacturers issued figures in varying sizes.

5-INCH SIZE. Lincoln Log. Miller and Slik Toy soldiers

4-1/2-INCH SIZE. Wilton.

4-INCH SIZE. Miller Nativity.

3-1/2- INCH SIZE. Ferdinand Gutmann, Molded Products, Playwood Plastics and Wilton.

3-1/4- INCH. Miller Nativity.

3-INCH. Described in the text and photograph captions as standard issue. All Nu/Faben, American Alloy, American Metal Toys, Auburn Rubber, Barclay (civilian, pot helmet and standard issue), Breslin, Cosmo, Grey Iron, H.B. Toys, Hill-John, Japanese Manufacturers, Jones, Manoil (composition, Happy Farm and My Ranch, Skinny Series, standard issue and 45/46 Series), Metal Cast, Miller (Farm) Noveltoy, Paul Paragine, Sheila, Soljertoy Tommy Toy, Trico.

2-3/4-INCH. Barclay Pod Foot, Historical miniatures, Manoil 500 Series.

2-1/2-INCH. Barclay Second Series, London Toy, Warren.

2-1/4 INCH. American Soldier Company, Barclay First Series, Beiser, Best Maid, Britains, Christies Toy and Metal Company, Grey Iron, Hahn, Japanese (medium size), Jones, Lincoln Log, Metal Cast, Minikin, Noveltoy, Soljertoy, Sonsco and William Feix. 2-1/4-inch size is classified as standard size in most collecting terminology. In Europe, 54 millimeter is used to describe this size as the first hollow cast figures produced in the United Kingdom by William Britain established the size criteria.

2-INCH SIZE. Arcade, Barclay Midi Series and Japanese (small).

1-3/4- INCH. Barclay 0-27 gauge, Jones/Cassidy 0-27 gauge railroad figures and Tru Craft.

OTHER SIZES. Some photographs show figures in unusually large sizes. In these cases, the size is indicated in the caption. Souvenir, Japanese composition and bisque and home cast figures vary in size between one and three quarter inch and five inch.

Types of Materials

It must be assumed that unless otherwise stated, figures are made of hollow cast lead. Aluminum, bisque, cast iron, clay, composition and solid are noted where appropriate.

Valuations Guide

The prices in each caption represent a comprehensive evaluation reflecting the supply and demand in the current toy soldier market. Many variables influence this market including: growth of collector interest, scarcity and condition, and geographical location. All these factors make it impossible to create an absolute price guide. These prices give the reader a guide to the current market values of Dimestore Toy Soldiers and Figures after taking these factors into consideration.

The value guidelines in this book reflect the price for a figure in "Good" (noticeable wear and paint loss from play) to "Very Good" (slight wear and paint loss) condition. Most toy soldiers offered in today's market fall within these two grades. Condition below "Good" would indicate considerable wear and paint loss and perhaps structural damage to the figure and has little value to a collector. Today, fewer and fewer toy soldiers are found in "Excellent" (no noticeable wear or paint loss, almost new) condition, and thus command a premium from collectors.

Figures considered to be unique or factory prototypes have not been given a value, the word unique has been designated in these cases.

Jones Father Time. A very rare figure. *Courtesy of the late Gus Hansen.* Unique item, value undetermined.

Part I

American Social History in Toys

1. Early American Manufacturers

American Soldier Company/Eureka, William Feix, Beiser, Theodore Hahn, Christies Metal and Toy Soldier Company, Soljertoy, Pearlytoy, McLoughlin, and Saint Louis Lead Soldier Company are most probably the best known American early or pre-dimestore manufacturers. Most, if not all, of these firms' products were influenced by the European manufacturers of the early twentieth century. Copying was commonplace resulting in the same figure being produced by more than one manufacturer. Most figures issued depict the uniforms of American or British troops from the early 1900s although cowboys and Indians also featured in some firms' products.

American Soldier Company. Eureka cavalry. The figure with the raised sword was possibly manufactured in England by William Britain. The round base cavalryman with side clips to base was intended to be used in an American Soldier Company boxed military game invented by Beiser. $25-$35 each.

American Soldier Company. New York Militia from Beiser military game. L to R: $15-$18 each.

American Soldier Company. New York Militia from Beiser military game, probably manufactured in England by William Britains. The underside of the base is marked "Made in Great Britain" instead of Made in England. $60-$80.

American Soldier Company. Eureka cavalry paint variants. $25-$35 each.

American Soldier Company probably made the four soldiers on the left and the one on the right. Figures five and six may have been made by Feix. $15-$18 each.

American Soldier Company. The ramrod and drummer figures are somewhat harder to obtain than the others. L to R: $20-$25, $20-$25, next six $15-$18 each.

Left to right: William Feix officers pointing and marching, Theodore Hahn, five variations and paint variants of officer with sword; similar figures were also produced by Christies Metal Toy Soldier Company. $15-$18 each.

Left to right: American Soldier Company/Beiser, three side drummers; unknown manufacturer; William Feix two side drummers and a fife player. $15-$18 each.

Left to right: American Soldier Company/Beiser, three paint variants; William Feix; unknown manufacturer. $15-$18 each.

Theodore Hahn. Five examples of American infantry circa 1910. $15-$18 each.

Theodore Hahn. Soldiers and sailors advancing with rifles. $10-$15 each.

Ferdinand Gutmann and Co., "Active Sammy on The Firing Line," jointed tin figure in three positions: charging, firing, and marching. $25-$35.

Opposite page;
Top left: Theodore Hahn. Three variations of kneeling firing soldier. $8-$10 each.

Top right: Left: unknown manufacturer. Soldier that shoots, possibly influenced by William Britain's soldiers. The barrel of the rifle is hollow; a tin tab (missing), is used to fire a missile through the soldier's rifle from the back. Middle: American Soldier Company/ Beiser, running at the slope. Right: Theodore Hahn charging figure. L to R: $35-$40, $15-$18, $8-$10.

Bottom left: The figure third left is Barclay first series, the other three are unknown early American manufacturers, though the Marine is possibly by Soljertoy. $10-$15 each.

Bottom right: Sailors. Left to right: American Soldier Company; Theodore Hahn; William Feix; Christies Metal Toy Soldier Company. $10-$15 each.

2. American Family

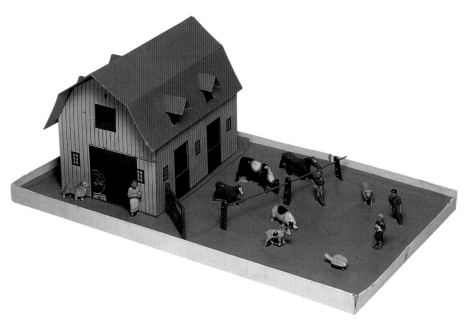

Grey Iron. "American Family" box lid.

Grey Iron. "American Family." Cardboard farm building, people, animals and fencing. $3,000-$4,000.

Grey Iron. "The American Family on the Farm." Calf, cow feeding, goat, horse, cow feeding (color variant), sheep, girl, man digging, gate, collie dog, goose, gate and post, farmer's wife, farmer and two pigs (paint variants). Pigs, calf, goat, sheep, dog and goose $4-$6 each. Dog, cow, horse $6-$9 each. People $10-$12 each. Gate and fences $15-$20.

Grey Iron. "The American Family at Home." Delivery boy, black female cook, boy with kite, milkman, black man digging, garage man, old couple sitting on bench, man with watering can, lady with basket, girl with skipping rope and collie dog.

Grey Iron. Left to right: "The American Family at Home:" boy with kite, collie dog, girl with skipping rope, old couple sitting on bench. "The American Family Travels:" lady wearing costume and man with traveling bag. L to R: $15-$20, $6-$8, $15-$20, Couple on bench, $10-$12 each, $10-$12, $10-$12.

Grey Iron. Left to right: "The American Family at Home:" garage man, rear and front view (two paint variants). Delivery boy, newspaper boy, from the "American Family Travels" series. Man with watering can and milkman also from the "At Home" series. L to R: $20-$25, $20-$25, $20-$25, $20-$25, $10-$12, $20-$25.

Grey Iron. "The American Family Travels." Back row (l-r): man in suit with bag (two paint variants), lady wearing costume, boy, girl, mailman, newspaper boy, preacher, old black man sitting on bench. Front row (l-r): conductor, engineer, black porter and policeman. L to R: (back row) $10-$12, $10-$12, $10-$12, $10-$12, $10-$12, $10-$12, $12-$15, $15-$20, $25-$30. (front row) $10-$12, $10-$12, $25-$30, $10-$12.

Grey Iron. "The American Family on the Ranch." Back row (l-r): calf, cowboy with rope, stallion, colt, burro. Front row (l-r): cowgirl in riding suit with crop, cowboy, cowboy riding bucking bronco, cowboy squatting, cowgirl on stallion, three ducks. A rooster and chickens complete this series (not illustrated). Calf, colt, burro, and ducks, $10-$12. Stallion, $15-$20. Cowboy with rope, girl in riding suit, boy cowboy and cowboy squatting, $20-$25. Cowgirl on stallion $35-$55. Cowboy on bucking bronco, $50-$60.

Grey Iron. "The American Family at the Beach." Back row (l-r): girl with large brim hat wearing slacks, boy, lifeguard in chair, boy with beach ball. Front row (l-r): man lying with newspaper over face, seated girl with sand pail, woman sitting in bathing suit, boy with life preserver, girl to catch ball, lifeboat. This set came boxed and complete with cabana (not illustrated). Girl with large hat, boy, boy with beach ball, seated girl with sand pail, woman sitting in bathing suit, boy with life preserver and girl to catch ball, $30-$40 each. Man lying with newspaper and lifeboat with seat, $40-$50. Lifeguard in chair, $60-$80.

Grey Iron. African-American figures. Man digging, female cook, old man seated on bench. L to R: $25-$30, $25-$30, and $35-$45.

Hubley. Black girl, lady, boy. L to R: $25-$30, $50-$60, $25-$30.

Manoil. "Darkie" eating watermelon from Happy Farm Series. $80-$100.

Grey Iron. Amish children at play in little red wagon, see-saw/teeter-totter, and swings. $20-$25 each.

Grey Iron. Amish children in sled, at school desks, and seated on bench. $20-$25 each.

Grey Iron. "Amish Family" group. $80-$100 set of 5.

Grey Iron. Amish salt and pepper shakers and egg timer. $20-$30 pair, $20-$25.

Barclay. Left to right: bridegroom and bride, minister walking, bridegroom and bride (paint variant);, minister holding hat (two paint variants), bridegroom and bride (paint variant). L to R: $30-$35 pair, $35-$40, $30-$35 pair, $15-$20, $15-$20, $30-$35 pair.

Unknown manufacturer. Cast iron Amish family. $80-$100 set of 4.

Barclay. Mini Series. Minister, bridegroom, and bride. $10-$15 each.

Japanese manufacturer. Bisque bridegroom and bride, sailor with bride. $15-$20, $25-$35.

Japanese manufacturer. Bride and bridegroom. $20-$25.

3. Railroad

Conductors by Barclay, John Hill and Company (made in England for Lionel Trains) and Trico Japanese (composition). L to R: $10-$15, $40-$50, $10-$15.

Jones/Cassidy. Boxed Railroad Miniatures set, American "O" gauge, containing young man with hands in pockets, young girl, newspaper boy, female railway passenger, engineer, signalman with lantern and colored porter. There is a space in this set for a conductor looking at watch (not illustrated). $80-$100 if complete.

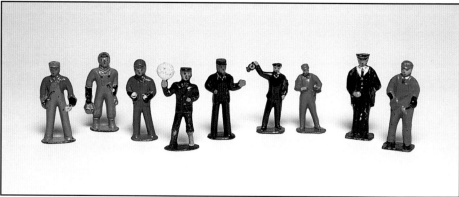

Left to right: Barclay, oiler, brakeman, engineer, peg-legged gateman, conductor from 0-27 Train Scale Series; Jones/Cassidy, signalman with lantern, engineer; Japanese manufacturer, conductor, brakeman. $10-$12 each except gateman $15-$20.

Left to right: Barclay engineer with oil can (three paint variants); Trico Japanese (composition) engineer with oil can; John Hill and Company, England, engineer with oil can made for Lionel Trains. $10-15, $10-$15, $10-$15, $10-$15, $40-$50.

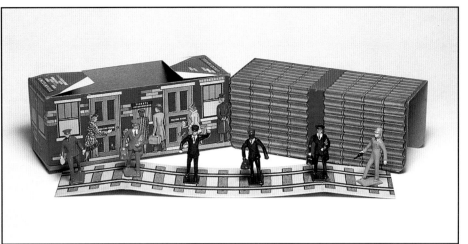

Lincoln Log. Cardboard railroad station display box and figures. $150-$200.

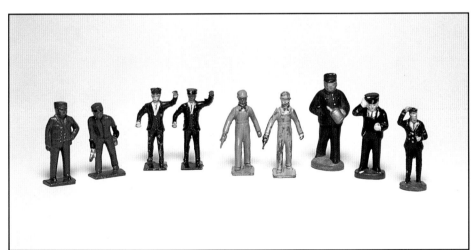

Left to right: Grey Iron, conductor and engineer from "The American Family Travels" set; Lincoln Log, pre- and post-war versions of conductor and engineer; three conductors of Japanese manufacture, two bisque and one Trico composition. $10-$15 each.

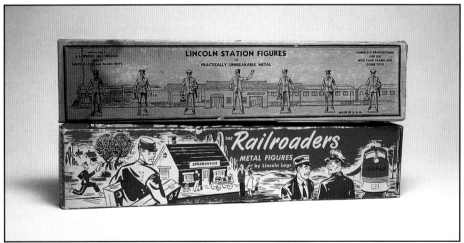

Lincoln Log. Two examples of box artwork for pre- and post- war railroad figures. $10-$15 each.

Tru Craft. Engineer, dining car attendant, and engineer with lantern. $10-$12 each.

Barclay. Left to right: lady with dog (two paint variants), gentleman with coat over arm (two paint variants), little girl with doll (two paint variants), boy (two paint variants), elderly lady and elderly man. $10-$15 each.

Barclay. Paint variant of elderly lady. $10-$15.

Barclay. Newsboys (two paint variants), shoe shine boy, girl in rocking chair and couple seated on bench. L to R: $10-$15, $10-$15, $20-$25, $10-$15, $25-$30 set.

Back row (l-r): Barclay 0-27 Scale "Train Series," man holding overcoat (two paint variants), woman, woman with dog, woman variation, woman with baby, hobo and newspaper boys (two paint variants). Front row (l-r): Japanese copies of Cassidy, man, woman, girl and two newspaper boys (paint variants); Barclay boy; unknown little boy; Barclay little girl. $10-$12 each except woman with dog $80-$100.

Left to right: Grey Iron, man in traveling suit (two paint variants), lady, boy, girl and newsboy from the "American Family Travels Set;" Lincoln Log, man carrying coat and bag (three paint variants). $10-$12 each except newsboy $12-$15.

Left to right: John Hill and Company, England, man with bag (made for Lionel Trains); Japanese manufacturer, two Trico male passengers, the second being a copy of Elastolin of Germany; John Hill and Company, female waving (made for Lionel Trains); Japanese manufacturer, two Trico female railway passengers. L to R: $40-$50, $10-$15, $10-$15, $40-$50, $10-$15, $10-$15.

Japanese manufacturer. Two bisque, two Trico compositions, three bisque (the second possibly Red Riding Hood), and four hollow cast lead railroad passengers. $10-$12 each.

Japanese manufacturer. Railroad figures in original box. $70-$80.

Left to right: Barclay, black porter with brush; John Hill and Company, England (made for Lionel Trains), black dining car attendant with foot stool and black porter; Japanese manufacturer, Trico copy of John Hill and Company's black porter; Barclay, black porter; Japanese manufacturer, Trico copy of John Hill and Company dining car attendant. The suitcase in the foreground is a salt and pepper shaker. L to R: $15-$20, $60-$80, $20-$30, $60-$80, $25-$30, $20-$30, $10-$12.

Left to right: Barclay, one rare and one standard issue black porter with luggage, from 0-27 Railroad Gauge Series; Japanese manufacturer, porter; Barclay, black porter; Jones/Cassidy, two porters; Japanese manufacturer, copy of a black porter with luggage; Barclay, dining car steward and porter paint variant in 0-27 Gauge Scale. $12-$15 each.

Left to right: Grey Iron, black porter; Lincoln Log, two paint variants of black porter. $10-$12 each.

Japanese manufacturer. Left to right: three unknown bisque black porters and two Trico examples in composition. $10-$12 each.

5. Farm

Manoil. "Happy Farm" series. Box lid artwork.

Manoil. "Happy Farm" series in original box. $300-$350.

Manoil. "Happy Farm" series. Cobbler mending shoe, blacksmith with wheel, man juggling/rolling barrel (three paint variants), farmer at water pump, man planting tree and shepherd boy with flute and crook. $20-$25 each, except man planting tree and shepherd boy, $30-$35.

Manoil. "Happy Farm" series. Young man and girl on bench, man carrying sack, blacksmith and anvil, farmer carrying pumpkin, watchman blowing out lantern, man chopping logs, and boy carrying wood. $20-$25 each.

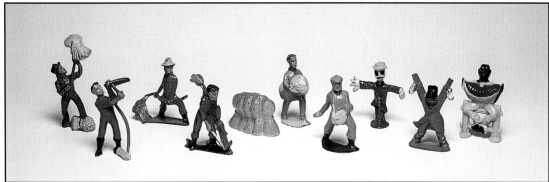

Manoil. "Happy Farm" series. Farmer pitching sheaves, sharpening scythe, cutting with scythe, farmer cutting corn, stack of wheat sheaves, man carrying sheaves, farmer sowing grain seeds, scarecrow with straw hat, scarecrow with top hat, black man sitting on fence eating watermelon. $20-$25 each, except scarecrow, $25-$30, and black man eating watermelon, $80-$100.

Manoil. "Happy Farm" series. Girl picking berries, woman with butter churn, woman lifting hen from nest, girl watering flowers, woman laying out washing on grass, lady holding baby, lady with pie, woman with broom, and school teacher with bell and cane. $20-$25 each, except school teacher, $30-$35.

Manoil. "Happy Farm" series. Carpenter carrying door, man with wheelbarrow, mason laying bricks, hod carrier with bricks, carpenter sawing, and carpenter with "T" square. $25-$30 each, except man with "T" square, $30-$35.

Manoil. Lady with pie, solid and all one color, available only from the Smithsonian Museum in Washington. $30-$35.

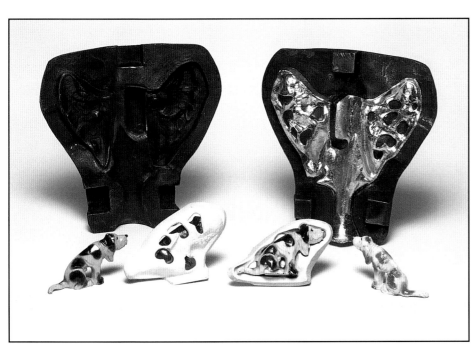

Manoil. Hound dog mold and paint spraying template. Three paint variants of a hound dog from the Happy Farm Series. Hounds, $15-$20 each.

Grey Iron. "Sunset Farm" in original box. Figures are solid, semi-round. *Photo courtesy Eccles Brothers Ltd*. $100-$120.

Top right: Farmers (l-r): Lincoln Log; home cast; American Metal Toys (three color variants). L to R: $15-$20, $10-$15, $15-$20, $15-$20, $15-$20.

Center right: Left to right: Auburn Rubber, farmer; Miller, farmer; Slikite; Judy Toy, farmer (two paint variants). L to R: $12-$15, $10-$12, $10-$12, $10-$12, $10-$12.

Bottom right: Farmer's wives (l-r): Lincoln Log; Judy Toy; American Metal Toys; Auburn Rubber; Miller. $10-$12 each.

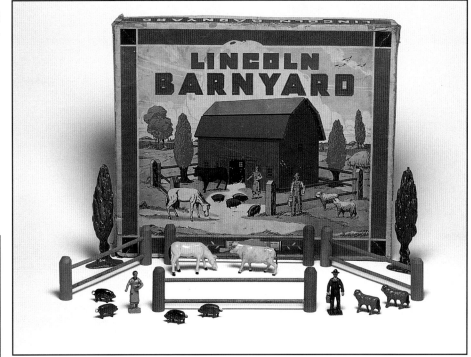

Lincoln Log. Barnyard figures and animals in original box. $125-$150.

Jones. Rare 54 mm. farmer with pitchfork and various farm animals, obviously influenced in design by Britains Ltd. Farmer $50-$60, animals $10-$30 each.

Lincoln Log. Barn, farmer, farmer's wife and tree. Barn $40-$50, figures $10-$12 each, tree $20-$25.

Miller. Farmer, farmer's wife, and farm animals. People $10-$12 each, animals $6-$10 each.

Minikin. Barnyard Series. Farm animals with original box. $80-$90.

Hobos (l-r): American Metal Toys; Barclay (small size); Trucraft; Barclay, plaster master figure; Molded Products. L to R: $15-$20, $10-$12, $8-$10, unique no price, $10-$12.

Horses. Left to right: Judy Toy; Grey Iron; composition horse; colt based on Auburn Rubber figures. $8-$10 each.

Horses. Back row (l-r): Barclay, work horse; American Metal Toys, Shire horse and quarter horse. Middle row (l-r): Lincoln Log; American Metal Toys, pony and lying colt; Barclay, walking horse. Front row (l-r): Barclay, feeding horse; Lincoln Log, feeding horse (large scale); American Metal Toys, walking colt. $10-$12 each.

Auburn Rubber. Farm wagon and team with a selection of horses, colts and ponies. Farm wagon, $35-$45; animals, $8-$10.

Cows. Back row (l-r): Barclay; Lincoln Log; Barclay, large size cow, which was intended to be part of nativity set, but never issued. Second row (l-r): Barclay, bull; American Metal Toys, calf; Barclay, feeding cow. Third row (l-r): American Metal Toys, lying cow, standing cow and lying cow (paint variant); Lincoln Log, feeding cow. Front row (l-r): Barclay, lying cow; American Metal Toys, calf (paint variant) and walking cow based on the Lincoln Log figure. $10-$12 each, Barclay nativity cow $20-$25.

Left and right: American Metal Toys, two sizes of donkey. Center: Auburn Rubber mule. $10-$12, each.

Cows. Back row (l-r): Auburn Rubber, three color variants; Judy Toy. Second row (l-r): Auburn Rubber, three paint variants, two different casting version of calves. Third row (l-r): unknown composition copies of Auburn Rubber cow and calf; Auburn Rubber, two paint variants. Front row (l-r): Auburn Rubber, second version calf; Molded Products, composition. $8-$10 each.

Sheep. Back row (l-r): American Metal Toys; Auburn Rubber, two versions. Second row (l-r): Barclay, ram; Judy Toy, sheep; Barclay, lying sheep; Auburn Rubber, second version. Front row (l-r): American Metal Toys; Barclay; Lincoln Log. $8-$10 each.

Pigs. Back row (l-r): American Metal Toys, three paint variants of a composition pig based on an Auburn Rubber figure; Auburn Rubber, pink and saddle-back pigs (paint variants). Second row (l-r): Judy Toy; Auburn Rubber, two versions. Third row (l-r): Molded Products; Auburn Rubber, three piglets and pig. Front row (l-r): Lincoln Log, two paint variants. $8-$10 each.

Left and right: All Nu, goats, two versions. Center: American Metal Toys, goat. $10-$12 each.

Back row (l-r): American Metal Toys, two hens and a rooster; Judy Toy turkey; Auburn Rubber and composition copy of Auburn Rubber turkeys. Middle row (l-r): two unknown composition hens; Auburn Rubber roosters (two paint variants); composition rooster; Auburn Rubber turkey (post-war version). Front row (l-r): Auburn Rubber hen; Judy Toy farmer's wife and cat; Auburn Rubber hen (paint variant). Poultry, $5-$6 each; Judy Toy farmer's wife, $8-$10; cats, $6-$8.

Back row (l-r): Judy Toy goose; American Metals Toys, goose; Wilton goose and duck; American Metal Toy duck. Middle row (l-r): Auburn Rubber duck and goose; Judy Toy duck; Auburn Rubber duck (paint variant); Wilton Mallard duck and red headed duck. American Metal Toys and Wilton $10-$12 each, others $4-$6 each.

Back row (l-r): American Metal Toys German Shepherd; Auburn Rubber large Collie; American Metal Toys German Shepherd (paint variant). Front row (l-r): American Metal Toys farm dog; Auburn Rubber small Collie; Judy Toy farm dog. $6-$12 each.

Back row (l-r): All Nu pointer dog; unknown Irish Setter, bronze plated; All Nu Dalmatian. Front row (l-r): black and brown Scottish Terriers; brown Scottish Terrier puppies; Dachshund; Wilton Dachshund. $10-$12 each.

American Metal Toys. Deer (six paint variants). Small size $25-$30, Large size $40-$50.

5. Indians

Back row and first three in center row: early American manufacturers, six paint variants of Indian brave with rifle issued by William Feix, Christies Metal and Toy Soldier Company, and Soljertoy. Center right and foreground: Japanese manufacturer. $8-$10 each.

Early American manufacturers. A number of the early pre-dimestore manufacturers seem to have issued versions of this Indian chief with tomahawk, among them, American Soldier Company, William Feix and Christies Metal and Toy Soldier Company. The figure was also produced by Reka in England. $10-$15 each.

Early American manufacturers. Back: (l-r): Beiser Shooting Game Indian with original brass hinged base made for them by Britains; William Feix, two paint variants of mounted brave with rifle (also issued as early Barclay). Foreground: Soljertoy mounted chief with rifle. L to R: Back: $20-$30, $15-$20, $15-$20. Foreground: $15-$20.

Barclay (l-r): three mounted Indian braves with rifles (two versions and two paint variants); mounted Indian brave (two feather version). L to R: $20-$25, $20-$25, $20-$25, $35-$40.

Barclay (l-r): mounted Indian braves firing rifles (two paint variants); Pod Foot Series, Indians firing rifles with shortened arms (two paint variants). L to R: $15-$20, $25-$35, $15-$20, $25-$30.

Left to right: Barclay, Indian chief with tomahawk and shield (pre- and post-war versions); Manoil, chief with tomahawk; Barclay Midi Series chief with tomahawk; Pod Foot chief with tomahawk (two paint variants). L to R: $12-$15, $12-$15, $80-$100, $40-$50, $8-$10, $8-$10.

Left to right: Barclay, Indian chief with knife; Manoil, four versions and paint variants of Indian chief with knife; Japanese manufacturer, dimestore Indian copy. L to R: $10-$12, $15-$20, $15-$20, $15-$20, $15-$20, $15-$20.

Left to right: Grey Iron, Indian chief with knife; H.B. Toys, brave with knife; unknown manufacturer, brave with knife; Molded Products, Indian chief; Wilton, cast iron brave. L to R: $10-$12, $10-$12, $10-$12, $8-$10, $25-$35.

Left to right: Barclay, Indian chief firing rifle, Indian brave with rifle, Pod Foot chief with rifle, and braves with spears (two paint variants); H.B. Toys; unknown manufacturer, brave with rifle. L to R: $40-$50, $10-$12, $15-$20, $8-$10, $8-$10, $8-$10, $10-$12, $10-$12.

Left to right: Barclay, Pod Foot Indian brave firing bow (two paint variants); Beton, brave with bow, later issued in plastic; Japanese manufacturer, chief firing bow. L to R: $15-$20, $15-$20, $15-$20, $20-$25.

Left to right: Barclay, Indian brave kneeling firing bow (two paint variants); H.B. Toys, kneeling Indian firing bow, Indian brave standing firing bow (two paint variants). L to R: $10-$12, $10-$12, $10-$12, $15-$20, $15-$20.

Left to right: American Metal Toys, brave kneeling firing rifle (two paint variants); Japanese manufacturer, brave firing rifle, foot on grass mound; unknown U.S. manufacturer, possibly a souvenir item; early Barclay brave; Barclay Midi Series brave firing a rifle. L to R: $35-$45, $35-$45, $15-$20, $35-$45, $10-$12, $45-$50.

Left to right: Grey Iron, Indian brave lying over horse firing rifle; American Metal Toys, brave on prancing horse; Grey Iron, mounted brave turned in saddle. L to R: $40-$50, $80-$100, $20-$25.

Left to right: H.B. Toys, chief kneeling with tomahawk; unknown manufacturer, chief kneeling with tomahawk; Grey Iron, Indian chief with tomahawk, brave with tomahawk (hand to brow), brave with raised tomahawk; Jones, Indian with warpaint and raised tomahawk. L to R: $12-$15, $12-$15, $10-$12, $15-$20, $50-$60, $100-$125.

Left to right: Soljertoy, Indian chief dancing with tomahawk; John Hill and Company, England, chief with tomahawk; Noveltoy, large chief walking with rifle at side; Japanese, composition Chief with tomahawk in moveable arm. L to R: $35-$45, $35-$45, $40-$50, $10-$12.

Left to right: early Barclay, First Series Chief firing bow (two paint variants); Japanese manufacturer; Lincoln Log; early Barclay, Second Series; Noveltoy, chiefs (two paint variants); Lincoln Log, brave firing bow. $10-$15 each.

Left to right: Lincoln Log, brave and chiefs with rifles (three paint variants); unknown manufacturer (possibly Noveltoy or early Jones), chief walking with rifle; Jones, chief advancing with rifle. L to R: $10-$12, $10-$12, $10-$12, $10-$12, $40-$50, $25-$30.

Lincoln Log. Indians and cowboys with cardboard teepee original box. (The original set contained fewer figures than are illustrated here.) $125-$150.

Lincoln Log or Noveltoy. Indian chiefs and braves mounted with rifles. $15-$20 each.

Lincoln Log. Boxed set containing Indians and cowboys. $125-$150.

Back row (l-r): Jones, mounted Indian chief firing bow backwards; Lincoln Log. Front: Japanese manufacturer, Indian brave and chief firing rifles. L to R (Back): $50-$60, $20-$25, (Front) $10-$15 each.

Japanese manufacturer. Original box for Indians. $10-$15.

Back row (l-r): Jones, Indian brave firing rifle; Japanese manufacturer, Indian brave and chief with spear (two paint variants); Lincoln Log; home-cast Indian chief with rifle. Front: Lincoln Log, crawling Indian (two paint variants). L to R: $25-$35, $10-$12, $10-$12, $10-$12, $8-$10, (Front) $10-$12, $10-$12.

Japanese manufacturer. Unusual Indian braves with two moveable arms carrying spear, tomahawk, and rifle. $35-$45 each.

Minikin, Japan. Mounted Indian chief with spear. $45-$55.

6. Cowboys

Left to right: Barclay, Cowboys walking (four paint variants), with tin brim hats; Noveltoy, cowboy with pistol. L to R: $10-$15, $10-$15, $10-$15, $10-$15, $35-$45.

Barclay. Cowboys with lassos (two paint variants and two casting versions). The second figure is the first version with loop to hold lasso above left hand. The other three are second version with flat bases. $10-$15 each.

Barclay. Cowboys with lassos, (three paint variants). Middle figure is post-war flat-based version (minus lasso) with hole through left hand. $12-$15 each.

Left to right: Barclay, Pod Foot cowboys with lassos, (three paint variants); Grey Iron, cowboy with lasso (missing). The third figure has original F.W. Woolworth store price tag attached. $10-$12 each.

Left to right: Barclay, cowboys firing pistols (three paint variants); H.B. Toys. cowboy. $10-$12 each.

Barclay. Pod Foot cowboys with pistols upraised (four paint variants). $10-$12 each.

Manoil. Cowboy with upraised pistol, (two oval hollow based paint variants and a hollow base version). $10-$15, $25-$30, $10-$15.

Left to right: Manoil, Grey Iron and Japanese manufacturer, surrendering cowboys. L to R: $10-$15, $60-$80, $10-$15.

Left to right: Grey Iron, cowboys drawing pistol (three casting versions and one paint variant); H.B. Toys, cowboy drawing pistol. L to R: $10-$12, $25-$30, $25-$30, $10-$12, $10-$12.

Left to right: American Metal Toys, kneeling cowboys with pistols (masked and unmasked versions with base); Barclay, Pod Foot cowboy firing pistol; Jones standing firing; H.B. Toys, kneeling cowboy firing pistol. L to R: $60-$70, $75-$100, $20-$25, $80-$100, $10-$12.

Grey Iron. Cowboy drawing pistol and black-face paint variant. L to R: $10-$12, $30-$35.

Left to right: Barclay, cowboy with two pistols; early Barclay second series with two pistols; unknown makers, two solid cast cowboys with rifle and pistol; Japanese manufacturer, cowboy with pistol and cowboy bandit. L to R: $50-$65, others $8-$10 each.

Left to right: Barclay, Pod Foot and Midi Series, two cowboys with rifles; Japanese manufacturer, kneeling firing rifle (two paint variants); John Hill and Company, England, cowboy; Japanese manufacturer, cowboy with rifle. L to R: $10-$12, $45-$55, $10-$15, $10-$15, $45-$55.

Left to right: Cosmo, cowboy firing two pistols; Japanese manufacturer; Molded Products (three paint variants); Japanese manufacturer, plaster composition with moveable pistol arm. L to R: $10-$15, $10-$15, $8-$10, $8-$10, $8-$10, $10-$12.

Left to right: Lincoln Log and Noveltoy, cowboys with ropes; Lincoln Log, cowboys with pistols (two paint variants); Noveltoy, cowboy with pistol; Jones cowboy with pistol. L to R: $10-$12, $10-$12, $15-$20, $15-$20, $15-$20, $15-$20.

William Feix. Mounted Cowboy with rope (two paint variants) with damage. $15-$20 each.

Barclay. Left to right: mounted cowboy with lassos (two versions of horse's tail), blue, yellow and red paint variants; two Pod Foot mounted versions, one with mask. L to R: $25-$35, $20-$35, $45-$55, $20-$25, $45-$50.

Left to right: early Barclay, mounted cowboy firing pistol, standard issue mounted figure with pistol, (three paint variants). L to R: $35-$45, $25-$30, $25-$30, $25-$30.

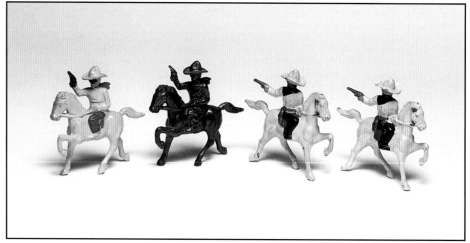

Left to right: All Nu/Faben, mounted cowboy with pistol and souvenir version in bronze finish; Barclay, mounted bandits firing pistols (two paint variants). L to R: $80-$100, $25-$35, $15-$25, $15-$25.

Left to right: American Metal Toys, mounted cowboy looking backwards with pistol; Barclay, mounted cowboy with pistol (paint variant); American Metal Toys, cowboy on rearing horse; H.B. Toys, cowboy firing backwards. L to R: $200-$250, $25-$30, $200-$250, $15-$20.

Left to right: Noveltoy, mounted cowboy with rifle; Barclay Pod Foot Series with pistol; Grey Iron, cowboy with rifle. L to R: $45-$55, $25-$30, $20-$30.

Left to right: Grey Iron, cowboy on bucking bronco; unknown manufacturers, souvenir rodeo cowboys. L to R: $35-$45, $20-$30, $20-$30.

Left and right: All Nu/Faben, cowboys on rearing horses. Center: unknown manufacturer souvenir cowboy with hat in air. L to R: $35-$50, $10-$15, $35-$50.

Left and right: All Nu/Faben, cowgirls on rearing horses. Center: Japanese manufacturer, cowgirl with whip. L to R: $35-$45, $20-$25, $35-$45.

Lincoln Log. Left to right: cowboy on rearing horse (two paint variants); cowboy waving hat in air (two paint variants). L to R: $40-$50, $40-$50, $20-$25, $20-$25.

Japanese manufacturer. Three rodeo cowboys with steers and horses. $30-$35 each.

Left to right: Japanese manufacturer, cowboy with rifle and pistol cowboy, possibly by Sonsco; unknown example based on the Britains' figure firing pistol backwards. L to R: $10-$12, $10-$12, $20-$25.

Japanese manufacturer. Boxed set of cowboys. $40-$50.

All Nu/Faben. Souvenir mounted cowboys. L to R: $25-$30, $40-$45.

Top left: Home-Cast. Tonto, Lone Ranger and Sheriff from original Home-Cast molds. *Courtesy Eccles Brothers Ltd.* Molds originally made by Allied Manufacturing. Current retail price.

Top right: Left to right: Jones; Lincoln Log, four versions of backwoodsmen. L to R: $100-$125, $15-$20, $15-$20, $15-$20, $15-$20.

Bottom right: Back: Lincoln Log, covered wagon with original box. Front: Barclay covered wagon. (Back) $60-$75, (Front) $35-$45.

Barclay. Covered wagon (paint variant). $40-$45.

Manoil. My Ranch Series. Not all of these items were produced from the molds that Manoil owned: Those that were are Brahma bull, bull, cow and feeding calf, foal, horse blankets, cowboy and cowgirl riding horses, entrance post for My Ranch Corral, prickly pear cactus and saguaro cactus. The camp fire scene, including the dancing couples, barrel cactus, and cowgirls and cowboys at the fence, was produced from the original molds by Eccles Brothers Limited, Iowa. Fence $8-$10, Brahma Bull, cow, calf, foal, $10-$15 each. Ranch entrance $35-$45. Horse blankets $40-$45, prickly pear cactus $20-$25, saguaro cactus, $35-$40, horse with cowboy or cowgirl rider $40-$45. All others are cast from original molds, current retail price.

Manoil. Mounted cowboy shooting (short and long pistol barrel version); mounted cowboy on trotting horse. $30-$45 each.

Manoil. My Ranch entrance and corral in original box. $75-$85.

7. Uniforms

Barclay. Hostess (three paint variants).
L to R: $175-$200, $60-$80, $35-$45.

All Nu. Girl's band saxophone player (paint variant). $80-$100.

All Nu. Girl's band. $80-$100 each.

All Nu. Majorettes. L to R: $100-$125, $400-$500.

Grey Iron. Gas pump attendant (paint variant). *See also the American Family Section.* $25-$35.

Left to right: Barclay, mailman; Barclay Mini Series, mailman; H.B., mailman; Grey Iron, Mailman; Lincoln Log, telegram messengers and pre- and post-war Western Union men. $10-12 each.

Firemen. Left to right: Barclay, prototype; H.B.; Barclay Mini Series. L to R: Unique $20-$25, $15-$20.

Barclay. Left to right: four versions of fireman with axe (two pre-war paint variants and two post-war flat based paint variants); fireman with hose. L to R: $20-$25, $20-$25, $20-$25, $20-$25, $25-$35.

Left to right: Barclay (two versions); Japanese manufacturer, bisque; Manoil (two versions); Manoil, prototype; Grey Iron, policeman. L to R: $10-$15, $10-$15, $10-$12, $15-$20, $15-$20, Unique, $15-$20.

Auburn Rubber. Post-war and pre-war motorcycle police versions. L to R: $15-$20, $35-$45.

Opposite page;
Top: Barclay. Motorcycle police (four versions). Each has a different cast engine. $35-$45 each.

Bottom: Left to right: Barclay Mini Series; Grey Iron; Lincoln Log (two pre- and one post-war version); Japanese manufacturer, bisque and lead policemen. L to R: $10-$12, $10-$12, $12-$15, $12-$15, $12-$15, $10-$12, $10-$12.

Barclay. Burglar and two detectives (two paint variants). L to R: $60-$80, $70 $90, $70 $90.

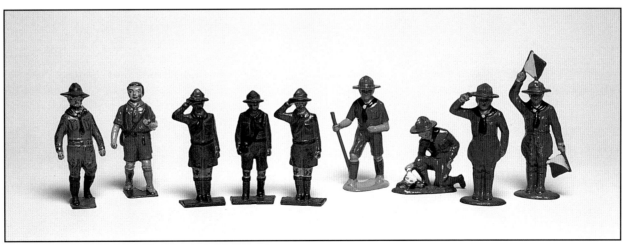

Boy Scouts. Left to right: Jones, two examples; Grey Iron, three examples; four Barclay examples, walking with stick, lighting fire, saluting, and with signal flags. L to R: $150-$200, $200-$250, $10-$15, $10-$15, $10-$15, $25-$30, $30-$40, $25-$30, $25-$30.

8. American Legion

Above: Barclay. American Legion paint variants and standard bearer. $200-$250 each. Standard bearer $350-$400.

Top right: Barclay, paint variants. $200-$250 each.

Bottom right: Barclay, paint variants and standard bearer. $200-$250 each. Standard bearer $350-$400.

9. Holidays

Minikin. Nativity set with original box. $100-$125.

Miller. Nativity set, one of a multitude of similar sets made for the dimestores in plaster or composition. $60-$80.

Barclay. Left: Santa Claus on tin skis; right: on lead skis; center: smaller size prototype on lead skis. L to R: $30-$40, unique, $30-$40.

Back row (l-r): Barclay, Santa on sleigh, Santa on sleigh pulled by reindeer; unknown manufacturer, Santa seated on sleigh (two paint variants). Front row (l-r): Japanese manufacturers, two miniature Santas on sleighs; Barclay, small size version. (Back) L to R: $200-$225, $150-$175, $40-$60, $40-$60. (Front) L to R: $15-$20, $25-$30, $20-$25.

Japanese manufacturer. Miniature Santa on sleigh. $35-$40.

Unknown manufacturer. Hollow cast Santa Claus. $60-$70.

Miller. Christmas scene with carol singers and dog. $25-$35.

Barclay. One horse open sleigh with original box. $65-$75.

Timpo-England. Uncle Mistletoe and Aunt Holly with donkey, pig, and rabbit. Made especially for Marshall Field's department store in Chicago. L to R: $60-$80, $150-$200, $150-$200, $60-$80, $60-$80.

Barclay. Winter scene showing examples of paint variants. *(See individual photographs)*

Barclay. Left to right: Skier with lead skis; two paint variants on tin skis. $20-$25 each.

Barclay. Man pulling sled with children, girl on sled, and man lying on sled. L to R: $35-$40, $15-$20, $15-$20.

Barclay. Back row (l-r): young man putting skates on girl sitting on bench; boy skater; two versions of girl skater. Front row (l-r): girl figure skater (two paint variants); male speed skater. (Back) L to R: $50-$75, $10-$12, $10-$12. (Front) $10-$15 each.

Grey Iron. Female skier, ice skater, and male skier. $30-$35 each.

Back row (l-r): Miller, plaster black cat, two sizes of witches. Front row (l-r): Tommy Toy, witch on broomstick; Miller, black cat and pumpkin heads; unknown manufacturer, witch on a broom stick with cauldron, possibly a souvenir item for Halloween. (Back) L to R: $40-$60, $65-$75, $50-$60. (Front) L to R: $125-$150, $40-$50, $35-$45, $50-$60.

Thanksgiving. Surprisingly enough manufacturers did not produce much for this holiday. Bisque and composition turkeys probably made in Japan. $10-$15 each.

Unknown manufacturer. Irishman with pig, probably a St. Patrick's Day souvenir, and metal cast Easter rabbit with Easter egg. L to R: $15-$20, $20-$30.

Part II

Military

10. Camp

Left to right: American Metal Toys, cook with pan; Barclay, cook with roast (egg timer); cook with roast standard issue. L to R: $80-$100, $65-$75, $15-$20.

Barclay. Soldier sitting with typewriter. $40-$60.

Left to right: Barclay, soldiers seated with rifles (two paint variants); Manoil, seated soldier with rifle across waist; American Metal Toys, seated soldiers with rifles (two paint variants), the second painted as enemy. L to R: $20-$25, $20-$25, $20-$25, $60-$80, $100-$125.

Back row (l-r): Manoil, soldier sitting and soldier eating. Barclay. Front row (l-r): soldier eating, typing figure at mess table. L to R: $25-$35, $25-$30, $25-$35, $15-$20.

Manoil. Man rolling barrel of apples (three paint variants). $30-$35 each.

Left and right: Manoil, cooks with ladles, second wearing a WW2 helmet. Center: soldiers seated peeling potatoes, (two versions). L to R: $35-$45, $25-$35, $25-$35, $65-$85.

Manoil. Paymaster. $100-$125.

Manoil. Soldiers seated writing letters, two versions, with and without cigarette. L to R: $35-$45, $55-$65, $55-$65.

Manoil. Seated banjo player and seated figure at wooden piano (folk art piano). L to R: $70-$80, $25-$35.

Manoil. Soldiers boxing. $60-$70 each.

Left to right: Manoil, photographers (thick and thin arm versions); Barclay, cameraman; All Nu, newsreel cameraman. L to R: $50-$60, $50-$60, $30-$35, $1200-$1500.

Camp scene showing Barclay and Manoil figures. The cardboard tent was made by Grey Iron. Tent $25-$35 (others priced in accompanying photographs).

11. Medical

Barclay. Left to right: kneeling nurse; nurse; three standard issue nurses (hair color variants); two Pod Foot nurses (hair color variants). L to R: $15-$20, $10-$15, $10-$15, $10-$15, $20-$25, $20-$25.

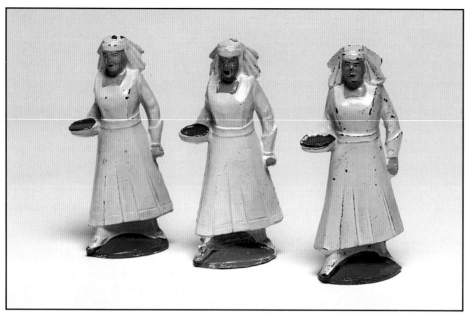

Manoil. Two hemmed skirt versions (paint variants) and a plain skirt version. L to R: $15-$20, $15-$20, 20-$25.

Left to right: Manoil, prototype with ether bottle (from an original mold produced by Eccles Brothers Ltd.); Grey Iron; American Metal Toys; Auburn Rubber; Tommy Toy. L to R: Unique, $10-$15, $50-$70, $10-$15, $145-$165.

Manoil. Casting version showing hemmed skirt and pointed veil. L to R: $15-$20, $20-$25.

Left: Unidentified, solid, possibly from a home cast mold. Right: Christies Metal and Soldier company nurse. L to R: $8-$10, $10-$15.

Barclay: Left to right: doctor; medical officer with Green Cross armband; medical officer with Red Cross armband; civilian doctor; doctor with wounded. L to R: $15-$20, $10-$15, $10-$15, $20-$25, $50-$70.

Manoil. Crawling medics with Green and Red Cross armbands. $35-$45 each.

Manoil. Doctor in khaki and two white coated paint variants. L to R: $25-$30, $15-$20, $15-$20.

Grey Iron. Medical officer and doctor with bag. L to R: $25-$30, $20-$25.

Left to right: American Metal Toys, white and khaki paint variants; Tommy Toy; Auburn Rubber. L to R: $40-$60, $60-$80, $100-$150, $15-$20.

Grey Iron. Left to right: sitting wounded; walking wounded; soldier helping walking wounded; and nurses with wounded. L to R: $45-$65, $25-$35, $175-$200, $100-$150.

Back row (l-r): Barclay, standard issue figure, Pod Foot soldiers on crutches (two paint variants), Pod Foot soldiers arm in sling (two paint variants); Manoil; Tommy Toy, arm in sling. Front: Barclay, seated wounded. L to R: $15-$20, $20-$25, $25-$30, $20-$25, $25-$30, $15-$20, $175-$200, (Front) $20-$25.

Left to right: American Metal Toys, two paint variants, second painted as enemy; unknown manufacturer, wounded. L to R: $200-$225, $250-$300, $25-$35.

Metal Cast. "The Sacrifice." World War One soldier falling wounded. $50-$60.

Barclay. Back row (l-r): stretcher party; open handed stretcher bearer; closed handed stretcher bearer; closed handed paint variant. Front row (l-r): two lying wounded, the right hand figure being a prototype. L to R: (Back) $45-$55, $45-$60, $10-$15, $10-$15, (Front) $10-$15, unique.

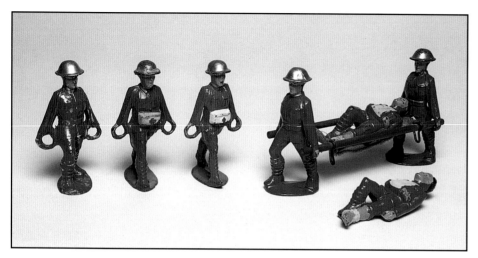

Manoil. Left to right: stretcher bearer without buttons on jacket; two versions with medical kit (the first with buttoned shirt); stretcher party and wounded. L to R: $10-$15, $50-$75, $15-$20, $55-$60, $10-$15.

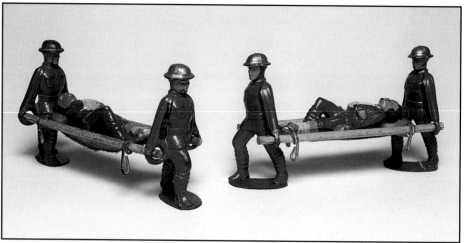

American Metal Toys, two paint variants, the second painted as enemy. L to R: $175-$200, $250-$300.

Manoil. 500 Series stretcher party. $200-$250.

Grey Iron. Stretcher party. $35-$45.

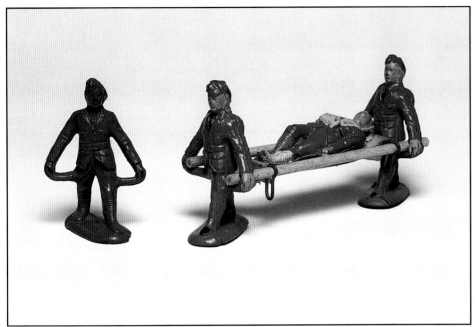

Auburn Rubber. Stretcher bearer paint variant, and stretcher party with wounded. L to R: $20-$15, $35-$45.

Miller. Stretcher party with nurse. L to R: $70-$80, $25-$35.

Playwood Plastics. Stretcher party. $20-$25.

Japanese manufacturer. Stretcher party in gas masks. $25-$30.

12. Officers

Barclay. Left to right: first series; second series saluting; clip-on helmet; short stride; glued helmet; cast helmet. L to R: $8-$10, $8-$12, $10-$15, $12-$16, $10-$15, $30-$55.

Barclay. Left to right: officer with orders (early and late paint variants); pot helmet; two Pod Foot color variants. L to R: $15-$20, $15-$20, $10-$15, $7-$10, $7-$10.

Manoil. Left to right: hollow base; standard issue; two paint variants of saluting general. L to R: $40-$60, $10-$15, $75-$100, $75-$100.

Left to right: Tommy Toy; American Alloy; All Nu; Metal Cast; H.B. Toys. L to R: $100-$150, $50-$80, $150-$300, $12-$15, $12-$18.

Grey Iron. First and second steel helmet versions and peak cap example. L to R: $10-$12, $12-$15, $10-$12.

Left to right: Auburn Rubber, first and second versions; American Metal Toys, two paint variants, The second painted as enemy; Jones, officer. L to R: $12-$15, $12-$15, $135-$200, $175-$250, $175-$225.

Left to right: Cosmo; John Hill and Company, England, thought to have been made for the U.S. market as part of a contract with the Lionel Train Company *(see also railroad section for other examples)*; Grey Iron, three Uncle Sam's Defenders. L to R: $25-$30, $35-$40, $5-$7, $5-$7. $5-$7.

Miller. General MacArthur. $40-$60.

Japanese manufacturer. Two bisque examples and four made in hollow cast lead. $10-$15 each.

13. Parade Figures

Barclay. Left to right: first version; second version; short stride; long stride, glued hat; long stride, clip hat, rifles at the slope. L to R: $10-$12, $7-$10, $10-$15, $10-$15.

Barclay. Left to right: tin helmet with pack; cast helmet with pack; pot helmet with pack; two Pod Foot paint variants, rifles at the slope. L to R: $15-$20, $20-$25, $20-$25, $7-$10, $7-$10.

Manoil. Hollow base and four variations of the standard issue figure with rifles at the slope. L to R: $25-$35, $10-$15, $20-$25, $35-$50, $10-$15.

Manoil. Left to right: Skinny Series with angled rifle; Skinny Series; 45/46 Series; 500 Series with rifles at the slope. L to R: $20-$25, $15-$20, $15-$20, $15-$20.

Manoil. Early and late version wearing overseas cap with rifles at the slope. L to R: $25-$35, $35-$50.

Grey Iron. Early and late versions in khaki and a legion paint variant with rifles at the slope. L to R: $10-$12, $10-$15, $20-$25.

Left to right, all with rifles at the slope: All Nu; Tommy Toy; Metal Cast; Paul Paragine. L to R: $150-$300, $75-$110, $10-$15, $40-$60.

Left to right, all with rifles at the slope: Grey Iron; Auburn Rubber; American Metal Toys, two examples, the second painted as enemy. L to R: $10-$15, $15-$20, $50-$80, $75-$100.

Left to right, all with rifles at the slope: American Alloy; H.B. Toys; Cosmo; unknown manufacturer. L to R: $30-$45, $12-$18, $30-$40, $15-$20.

Left to right, all with rifles at the slope: Miller; John Hill and Company, England, thought to have been made only for the U.S. market *(see also Officers and Railroad section)*; Japanese manufacturer, two ceramic and three hollow cast. L to R: $15-$25, $55-$65, $10-$12, $10-$12, $10-$12, $10-$12, $10-$12.

Left to right, all with rifles at the slope: Japanese manufacture, Best Maid; unknown manufacturer, ceramic; Jones (two versions); uncertain origin, possibly Early Jones. L to R: $8-$10, $8-$10, $20-$25, $20-$25, $35-$40.

Left to right: Jones, two versions with slung rifles and two at the slope; Lincoln Log, three versions, the third being from a home cast mold; two Noveltoy with rifles at the slope. L to R: $18-$20, $30-$40, $18-$20, $18-$20, $8-$10, $8-$10, $6-$8, $15-$20, $15-$20.

Barclay with slung rifles. Left to right: cast hat; pot helmet; large size prototype; Pod Foot; Midi Series. L to R: $15-$20, $15-$20, unique, $7-$10, $45-$70.

Manoil. Left to right: slung rifle; rifle at angle; sentry; 500 Series in pancho. L to R: $10-$15, $80-$125, $45-$60, $25-$30.

Left to right: Molded Products, three versions; Playwood Plastics, two versions with slung rifles. L to R: $7-$10, $7-$10, $7-$10, $8-$10, $8-$10.

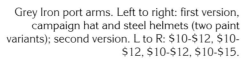

Playwood Plastics. Overseas hats and slung rifles, two versions. $10-$15 each.

Barclay port arms. Left to right: tin helmet; cast helmet; pot helmet. L to R: $10-$15, $15-$20, $15-$20.

Grey Iron port arms. Left to right: first version, campaign hat and steel helmets (two paint variants); second version. L to R: $10-$12, $10-$12, $10-$12, $10-$15.

Left to right: Auburn Rubber, two versions; Jones Casting; early Jones. All port arms. L to R: $20-$30, $20-$30, $10-$15, $30-$35.

Manoil present arms. Four paint variants. $15-$20 each.

Left to right, all at attention: Barclay, second series slope arms, tin helmet, pot helmet; Tommy Toy; Jones. L to R: $10-$15, $15-$20, $15-$20 $125-$175, $175-$225.

Left to right: Jones (two paint variants) at ease; Arcade; Grey Iron, Uncle Sam's Defenders; Japanese manufacturer bisque at attention. L to R: $18-$20, $18-$20, $6-$8, $3-$45, $3-$5, $8-$10.

London Toy. Canadian infantry at attention. $10-$15 each.

Manoil. Left to right: 45/46 Series, standing at the ready; 500 Series standing at the ready. $15-$20 each.

Left to right: Barclay, Grey Iron, and Barclay prototype sentries. L to R: $12-$18, $15-$20, unique.

Miller and Barclay sentries with dogs. L to R: $30-$50, $35-$50.

14. Musicians

Drum majors. Left to right: Barclay, short stride, long stride, white pot helmet; Grey Iron; two Japanese manufacturers. L to R: $12-$18, $12-$18, $35-$50, $15-$20, $10-$12, $12-$15.

Barclay buglers. Left to right: first version; second version; short stride; long stride; long stride (paint variation); white pot helmet; three Pod Foot (paint variations); Midi. L to R: $12-$18, $8-$12, $12-$18, $10-$15, $30-$35, $7-$10, $7-$10, $7-$10 $45-$70.

Manoil buglers. Left to right: Hollow base, two standard issue versions. L to R: $40-$60, $15-$20, $15-$20.s

Auburn Rubber buglers. Five variations. $10-$15 each.

Buglers. Left to right: All Nu; American Metal; Grey Iron, first version; Grey Iron, second version; two H.B. Toys versions. L to R: $500-$750, $85-$135, $15-$20, $15-$20, $12-$18, $15-$20.

Barclay drummers. Left to right: first version (two paint variants); short stride (two paint variants); long stride; white pot helmet. L to R: $12-$18, $12-$18, $12-$18, $12-$18, $12-$15, $30-$35.

Manoil. Hollow base and two versions of drummers. L to R: $40-$60, $15-$20, $25-$40.

Grey Iron. First and second version drummers. $15-$20 each.

Japanese manufacturer drummers. The first is bisque and the others hollow cast lead. L to R: $10-$12, $12-$15, $12-$15, $12-$15.

Barclay. French horn player and white pot helmet sousaphone player. L to R: $12-$18, $30-$35.

Left to right: Barclay, fifer, white pot helmet clarinet player; Japanese manufacture, fifer with moveable arms; Barclay, second series fifer; two Japanese manufacturer, clarinet players. L to R: $12-$18, $30-$35, $8-$10, $8-$12, $8-$10, $8-$10.

Barclay. White pot helmet bandsmen. L to R: $35-$50, $30-$35, $30-$35, $30-$35, $30-$35.

Japanese manufacturer. U.S. Military Band. $8-$10 each.

Japanese manufacturers. Bandsmen showing the variations in style of design. $10-$12 each.

Japanese manufacturer. U.S. Military Band. $120 set.

15. Flagbearers

Barclay. Left to right: second series marching; first series; second series at attention; short stride; long stride; cast helmet; Navy; American Legion; pot helmet; two Pod Foot (paint variants). L to R: $10-$12, $10-$15, $8-$12, $15-$20, $10-$15, $15-$20, $15-$25, $350-$400, $15-$20, $7-$10, $7-$10.

Left to right: Grey Iron; American Metal; Metal Cast; Metal Cast casting; Japanese manufacturer dimestore copy. L to R: $15-$20, $95-$145, $40-$60, $12-$15, $45-$55.

Manoil. Left to right: hollow base; two pre-war versions; Skinny Series; 45/46 Series; 500 Series. L to R: $45-$70, $15-$20, $20-$25.

Left to right: Auburn Rubber; three Molded Products versions; Playwood Plastic. L to R: $60-$80-$8-$12, $8-$12, $8-$12, $12-$15.

Top left: Left to right: Breslin with unidentifiable flag; copy of Trico Japanese composition in lead; two Grey Iron, the second with rare Cuban Flag. L to R: $40-$50, $30-$40, $200-$300, $400-$500.

Bottom left: Left to right: Grey Iron, Uncle Sam's Defender Series; Japanese bisque; Japanese hollow cast; Japanese Dimestore copy. L to R: $5-$7, $18-$20, $10-$12, $30-$40.

Top right: Miller. Two versions. L to R: $20-$30, $15-$25.

Jones. Left to right: Naval Academy; U.S. Army World War One; U.S. Army World War Two. L to R: $35-$45, $200-$250, $60-$80.

Metal Cast. Flag bearer. $40-$60.

Left to right: American Soldier Company; Beiser or Eureka, possibly made to be incorporated in the Beiser shooting games retailed by Britains in England. $15-$25 each.

Historical Miniatures. Colonial flag bearer. $60-$80.

16. Signal Flags

Signalmen. Left to right: Barclay; Manoil, hollow base; Manoil, standard issue; Grey Iron, Naval. L to R: $15-$20, $25-$35, $20-$30, $20-$25.

Opposite page: Signalmen. Left to right: Barclay; All Nu; Grey Iron; Auburn Rubber, second version and first version. L to R: $15-$20, $900-$1400, $25-$30, $20-$40, $150-$180.

Barclay. Boy Scout. $20-$30.

Lincoln Log. VERY RARE figure with signal flags. 5-1/2 inch high. $175-$250.

17. Cavalry

Barclay. Back row (l-r): Four early first and second series, 2-1/4 inch cavalry. Front row (l-r): two small-scale cavalry. The orange and blue tunics are possibly meant to depict foreign troops. L to R: (Back) $26-$40, $26-$40, $26-$40, $28-$42. (Front) $12-$18, $12-$18.

Barclay. Cavalry officers (four standard issue paint variants). $25-$35 each.

Barclay. Four uniform paint variants. The red tunic represents New York Militia, the others depict foreign troops. L to R: $20-$25, $25-$35, $25-$35, $25-$35.

Grey Iron. Back row (l-r): campaign hats first version; second version of cavalry (two paint variants); peak cap, second version. Front row (l-r): peak cap, first version; Legion officer on white horse. L to R: (Back) $20-$25, $20-$25, $20-$25, $25-$30. (Front) $20-$25, $30-$45.

Barclay. Left to right: two intermediate versions; Pod Foot; a prototype. L to R: $35-$50, $35-$50, $50-$75, unique.

Left to right: Molded Products, two examples; Lincoln Log, two examples. L to R: $10-$15, $10-$15, $10-$12, $10-$12.

Left to right: Noveltoy; Jones (two paint variants). L to R: $50-$60, $35-$45, $35-$45.

Auburn Rubber. Paint variants. $25-$40 each.

Metal Cast. Left to right: three 54 mm scale; a standard issue peak cap version. L to R: $25-$30, $25-$30, $20-$25, $20-$25.

Japanese manufacturers. Two ceramic and two hollow cast lead. L to R: $12-$18, $10-$15, $10-$15, $10-$15.

Left to right: early, pre-Dimestore manufacture, possibly Hahn, three versions; Soljertoy. $12-$15 each.

18. Communications

Left to right: Barclay, Midi and two standard size paint variants with field telephones; Barclay, two paint variants with field telephone and antenna; Manoil, seated with phone at map table; American Metal Toys, seated with phone. L to R: $45-$70, $10-$15, $10-$15, $25-$50, $25-$50, $20-$30, $55-$85.

Manoil. Left to right: radio operator; lineman and telegraph poles (two versions); lying radio operator. L to R: $35-$45, $50-$75, $50-$75. (Front) $30-$40.

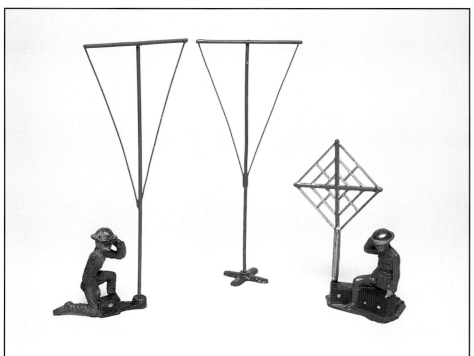

Left: Grey Iron, wireless operator with antenna. Right: Barclay, wireless operator. L to R: $25-$50, Set $150-$225, $20-$30.

Barclay. Left to right: pigeon handler; raft with two men. L to R: $20-$25, $35-$50.

Left to right: American Alloy, two versions with walkie-talkies; Miller; Slik Toy. L to R: $50-$75, $50-$75, $15-$25, $15-$20.

19. Firing Figures

Barclay. Left to right: second series, two versions; short stride; long stride; long stride cast helmet; pot helmet; Pod Foot (two paint variants). L to R: $12-$18, $8-$12, $10-$15, $10-$15, $10-$15, $15-$20, $7-$10, $7-$10.

Left to right: Manoil, folding rifle, standard issue figure, 500 Series; American Metal Toys; American Metal Toys, enemy; All Nu. L to R: $190-$275. $10-$15, $15-$20, $50-$75, $75-$100, $100-$250.

Left to right: Auburn Rubber;. Manoil, two 500 Series, 45/46 Series, and Skinny Series. L to R: $20-$35, $20-$25, $20-$25, $20-$25, $20-$25.S

Manoil, all with automatic weapons. Left to right: 500 Series; two 45/46 Series, one with white helmet; Skinny Series. L to R: $15-$20, $15-$20, $18-$22, $15-$20.

Left to right: Japanese manufacturer; Arcade; Jones. L to R: $18-$20, $8-$10, $20-$25.

Manoil. Left to right: 500 Series; standard versions thick rifle; folding rifle version; standard version thin rifle. L to R: $20-$25, $10-$15, $165-$250, $10-$15.

Barclay. Pod Foot (three paint variants). L to R: $7-$10, $7-$10, $50-$75.

Manoil. A close-up showing both standing and kneeling folding rifle figures. L to R: $190-$275, $165-$250.

Left to right: Paul Paragine; unknown manufacturer; Japanese; Jones; Arcade. L to R: $50-$70, $25-$30, $15-$20, $35-$45, $7-$10.

Barclay, first series. Left to right: short stride; clip on helmet; glued helmet (two versions), the second has thin rifle; long stride; pot helmet. L to R: $10-$15, $10-$15, $10-$15, $10-$15, $10-$15, $10-$15, $25-$35.

Left to right: Grey Iron; Auburn Rubber (two paint variants); American Metal Toys, thin and thick rifle versions, and enemy paint variant. L to R: $15-$20, $15-$20, $15-$20, $60-$90, $75-$110, $75-$125.

Left to right: Tommy Toy; Playwood Plastic; H.B. Toys; Lincoln Log. L to R: $120-$170, $6-$10, $12-$18, $100-$150.

Left to right: Manoil camouflage sharpshooter; Barclay (two paint variants). L to R: $15-$25, $10-$15, $10-$15.

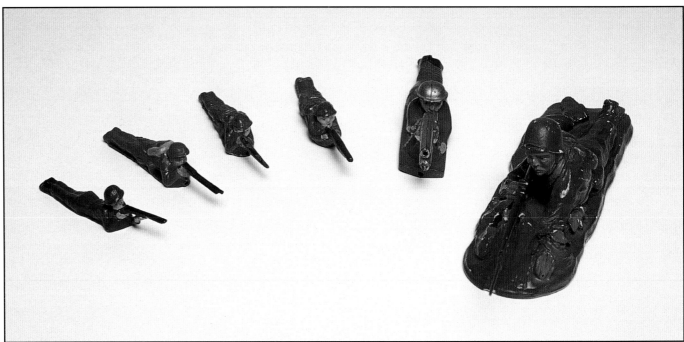

Left to right: Japanese manufacturer, four figures; American Metal Toys; Miller. L to R: $10-$15, $10-$15, $10-$15, $10-$15, $60-$90, $15-$25.

Auburn Rubber. "Tanker." $20-$30.

Barclay. Left and right: two examples of soldier firing from behind a wall. Center: a prototype pill box group. L to R: $35-$50, unique, $35-$50.

20. Machine Gunners

Barclay lying machine gunners. Left to right: tin helmet; cast helmet; cast helmet variant; tin helmet paint variant; pot helmet. L to R: $10-$15, $10-$15, $10-$15, $10-$15, $15-$20.

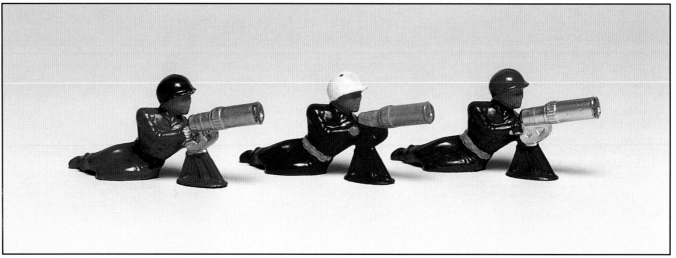

Barclay. Pod Foot lying machine gunners (three paint variants). L to R: $7-$10, $50-$75, $7-$10.

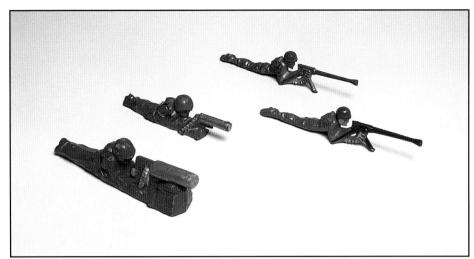

Manoil clay/composition lying machine gunners produced during World War Two when lead was in short supply. Left to right: 500 Series, Skinny Series, unpainted casting; painted Skinny Series. L to R: $30-$45, $15-$20, $15-$20, $35-$50.

Manoil. Five pre-war versions of lying machine gunners. L to R: $15-$20, $10-$15, $25-$30, $10-$15, $10-$15.

Manoil. Lying on vehicle with machine gun. $35-$45.

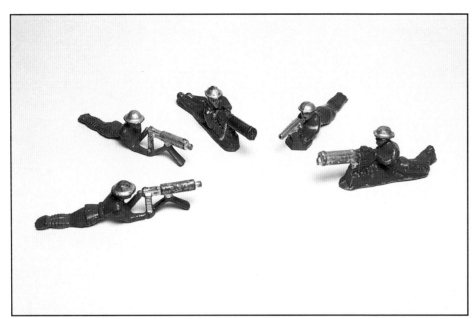

Lying machine gunners. Left to right: Grey Iron (two paint variants); Breslin; American Metal Toys; American Alloy. L to R: $10-$15, $10-$15, $20-$25, $65-$90, $40-$60.

Lying machine gunners. Left to right: Playwood Plastics, two versions; Molded Products, two versions. L to R: $8-12, $9-$12, $7-$10, $7-$10.

Lying machine gunners. Left to right: American Metal Toys; American Metal Toys/Jones (German); Breslin; American Alloy. L to R: $65-$90, $80-$125, $30-$35, $40-$60.

Lying machine gunners. Left to right: Lincoln Log, three versions; Jones; Lincoln Log, home cast; Japanese bisque; Molded Products. L to R: $8-$10, $8-$10, $8-$10, $15-$18, $6-$8, $15-$18, $7-$10.

Japanese manufactured hollow cast. Two examples of lying machine gunners and bullet feeders. L to R: $12-$15, $15-$20, $10-$12, $12-$15.

Barclay. Left to right: first series; second series; short stride; short stride (paint variant); cast helmet; seated, rather than kneeling; tin helmet. L to R: $10-$15, $8-$10, $8-$10, $15-$20, $10-$25, $10-$15.

Manoil. Left to right: three pre war versions; Skinny Series; Skinny Series paint your own version; 500 Series; Composition/clay. (Left) $15-$20 each. (Center) Top $15-$20. (Bottom) $35-$40. (Right Top) $15-$20, $25-$40.

American Metal Toys. Left side: two versions (three paint variants, top painted as enemy); Right side, bottom to top: All Nu; Metal Cast; Playwood Plastics. (Left) Top to Bottom $60-$80, $35-$65, $75-$100 with number on pocket. (Right) Top to Bottom $6-$10.

Left: Auburn Rubber, two versions. Center: Grey Iron (front); H.B. Toys (back); Right: two Grey Iron, one painted as Legion. (Left) $8-$10 each. (Front Center) $8-$10. (Back Center) $12-$18. (Right) $10-$15 each.

Left to right: Grey Iron, Uncle Sam's Defenders; Jones; Jones Russian sailor; unknown bisque; Slik Toy. L to R $5-$7, $20-$25, $25-$30, $15-$20, $20-$30.

Machine gunners with loaders. Left to right: Manoil, two versions; Breslin; Playwood Plastics. L to R: $20-$25, $20-$25, $6-$10, $6-$10.

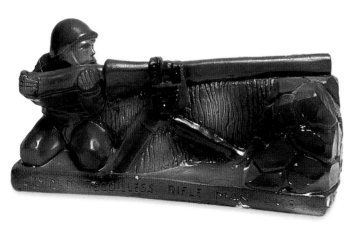

Slik Toy. Recoil rifle. $25-$30.

21. Grenade Throwers

Barclay. Left to right: second series; short stride; large size; long stride; long stride cast helmet; Pod Foot (two paint variants). L to R: $8-$12, $10-$15, $300-$450, $15-$20, $20-$25, $7-$10, $7-$10.

Left to right: Manoil; Manoil 500 Series; Auburn Rubber; two American Metal Toys, the second painted as enemy. L to R: $35-$50, $25-$30, $10-$15, $70-$80, $90-$150.

Left to right: Jones; Tommy Toy; All Nu; American Alloy; Molded Products. L to R: $100-$125, $900-$1400, $75-$100, $5-$8.

Left to right: Miller; Arcade; Slik Toy. L to R: $15-$25, $5-$7, $10-$15.

22. Advancing and Charging

Barclay. Left to right: second series; first series; advancing; glued hat; clip on hat; standard issue charging figure (two paint variants). L to R: $8-$12, $7-$10, $10-$15, $10-$15, $10-$20, $15-$20.

Barclay. Left to right: advancing (two paint variants); RARE advancing version with rifle sling; RARE large size charging; tin helmet; charging cast helmet (two paint variants). L to R: $15-$18, $15-$18, $450-$650, $300-$450, $25-$30, $15-$20, $15-$20.

Barclay. Left to right: Pod Foot (three paint variants); two advancing with head turned; two examples of Midi Series. L to R: $50-$75, $7-$10, $7-$10, $7-$10, $7-$10, $45-$70, $45-$70.

Left to right: Manoil, two versions advancing with automatic weapons; Auburn Rubber, two versions advancing with rifles; Playwood Plastics, two automatic weapon versions. L to R: $20-$30, $10-$15, $20-$30, $10-$15, $8-$12, $8-$12.

Manoil. Left to right: two versions of the charging figure with bayonet; charging without bayonet; advancing with bayonet. L to R: $20-$30, $20-$30, $25-$35, $25-$45.

Left to right: H.B. Toys advancing; American Alloy; Tommy Toy advancing; All Nu, charging with bayonet; Playwood Plastics. L to R: $12-$18, $125-$175, $100-$150, $500-$750, $8-$12.

Advancing. Left to right: unknown manufacture; American Metal Toys (two paint variants, the second painted as enemy); American Alloy. L to R: $50-$70, $150-$200, $200-$250, $40-$60.

Advancing or charging wearing gas masks. Left to right: Barclay; Breslin; Barclay variation; Barclay cast helmet; All Nu; Manoil; uncertain origin, possibly Manoil. L to R: (Back) $10-$15, $20-$25, $20-$25, $15-$20, unique. L to R: (Front) $10-$15, $900-$1400.

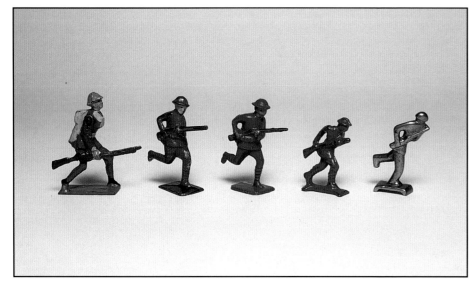

Advancing or charging. Left to right: Jones; Japanese manufacturer; Lincoln Log; Grey Iron, Uncle Sam's Defenders; Arcade. L to R: $30-$35, $8-$10, $8-$10, $5-$7, $3-$5.

Grey Iron. Left to right: early version in campaign hat; late version; Legion in helmet; early helmet version. L to R: $10-$12, $12-$15, $20-$25, $10-$12.

Advancing in crouching positions. Left to right: Barclay. Manoil, and Breslin. L to R: $15-$20, $35-$50, $25-$30.

Crawling with pistols. Back row (l-r): Barclay; Pod Foot (two paint variants); Manoil. Front row (l-r): Grey Iron; Jones (two color variants). L to R (Back) $7-$10, $7-$10, $25-$35, $15-$20. (Front) $15-$20 each.

Crawling. Back row (l-r): Metal Cast casting; Manoil, two versions; American Metal Toys. Front row (l-r): Barclay (two versions the second falling with rifle); Auburn Rubber. L to R (Back) $15-$20, $90-$135, $25-$40, $70-$100. L to R (Front) $10-$15, $20-$30, $25-$45.

Barclay. Left to right: tin helmet; cast helmet with machine guns; tommy gunner; Pod Foot (two paint variants); a Midi Series with tommy gun. L to R: $10-$15, $18-$25, $15-$20, $7-$10, $7-$10, $45-$70.

With automatic weapons. Left to right: All Nu; Tommy Toy; Jones (two color variants). L to R: $900-$1400, $150-$200, $15-$18, $20-$25.

Left to right: Miller, kneeling and standing with automatic weapons; Slikite; Slik Toy. L to R: $30-$40, $25-$35, $10-$15, $20-$25.

Miller. Left to right: advancing; soldier in foxhole; charging. $15-$25 each.

Left to right: Barclay, walking with rifle raised; Manoil, clubbing with rifle; Grey Iron (two versions) stabbing downward. L to R: $10-$15, $30-$40, $15-$20, $20-$25.

American Metal Toys. Stabbing with bayonet, the first painted as enemy. L to R: $100-$150, $150-$200.

Barclay. Left to right: bayoneting, tin hat and cast hat; clubbing with rifle, cast hat and tin hat. L to R: $30-$45, $150-$250, $50-$75, $25-$40.

23. Flame Throwers and Bazookas

Barclay. Left to right: Pod Foot (three paint variants); Midi Series. L to R: $50-$75, $7-$10, $45-$70, $7-$10.

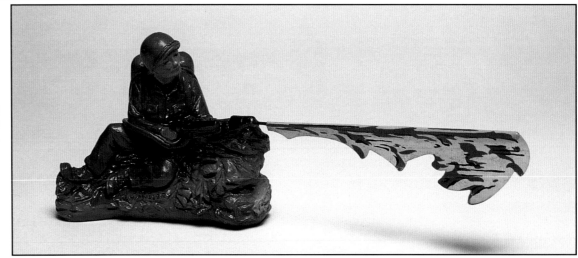

Miller. The flame section is a wire rod with composition "fire" which slots into the figure. $30-$50.

Miller. Lying and kneeling versions with detachable weapons. L to R: $25-$35, $20-$35.

Barclay. Pod Foot (four paint variants) and a kneeling Midi Series. L to R (Back) $7-$10 each. (Front) $45-$70.

Back row (l-r): Manoil, 45/46 Series loader and bazooka operator; American Alloy. Front row (l-r): Manoil, 500 Series, (two paint variants). L to R (Back) $20-$25, $20-$25, $60-$80. L to R (Front) $15-$20 each.s

Manoil. 45/46 Series loader, bazooka operator with white helmet, and a paint variant. $20-$25 each.

24. Gas Mask Figures

Left to right: Barclay, two charging (paint variant), and a cast helmet version; Manoil, two grenade thrower versions. L to R: $10-$15, $10-$15, $15-$20, $15-$20, $15-$20.

Barclay. Howitzer and men (three paint variants). $15-$20 each.

Gas masked and carrying pistols. Left to right: Barclay, (two paint variants); Tommy Toy; American Alloy; Metal Cast, World War I; Metal Cast, World War II. L to R: $15-$20, $15-$20, $175-$250, $80-$110, $40-$60, $15-$20.

Left four with rifles: Manoil; Breslin; uncertain origin, possibly Manoil; All Nu. Center two with flare guns: Manoil; Manoil 500 Series. Right two with bayonets: American Metal Toys, the first painted as enemy. L to R: $15-$20, $15-$20, unique, $900-$1400, $15-$20, $20-$25, $100-$150, $150-$200.

Grenade Throwers. Left to right: Manoil (two versions); Breslin; Metal Cast. L to R: $15-$20, $15-$20, $15-$20, $40-$60.

Left to right: Best Maid, Japan, stretcher party and individual stretcher bearer; Jones, Irish machine gunner. L to R (Back) $50-$55, $10-$15. (Front) $20-$25.

Left to right: Molded Products, (three versions) with pistol and a grenade thrower; Playwood Plastics (two versions) with flare gun. L to R: $7-$12, $7-$12, $7-$12, $5-$8, $6-$10, $6-$10.

25. Anti-Aircraft

Barclay anti-aircraft guns and gunners. Left to right: two tin helmet paint variants; cast helmet (two paint variants); pot helmet. L to R: $10-$15, $10-$15, $12-$18, $15-$20, $12-$18.

Barclay anti-aircraft gun and gunners. Pod Foot Series (front and rear views). $15-$20 each.

Barclay anti-aircraft gun. Left to right: two cast helmet (paint variants); three pot helmet (paint variants). L to R: $15-$20, $15-$20, $25-$35, $25-$35, $25-$35.

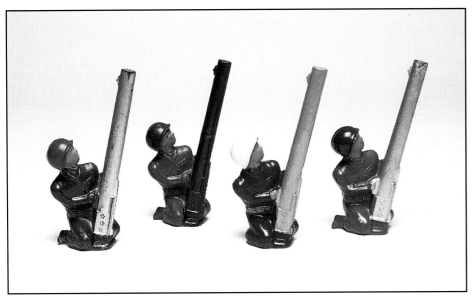

Barclay. Pod Foot. anti-aircraft guns (four paint variants). L to R: $7-$10, $7-$10, $50-$75, $7-$10.

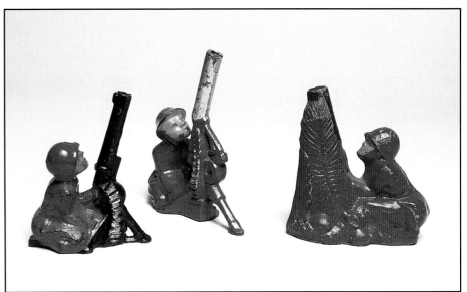

Manoil anti-aircraft guns. Left to right: 500 Series; standard issue; clay composition version. L to R: $25-$30, $12-$18, $25-$40.

Manoil. Left to right: anti-aircraft gun with range finder; anti-aircraft guns and gunners (two versions). L to R: $10-$15, $10-$15, $12-$18.

Anti-aircraft guns. Left to right: American Metal Toys, (two versions, second painted as enemy); All Nu. L to R: $45-$65, $60-$80, $350-$500.

Left to right: Auburn Rubber, anti-aircraft gun; Lewis, machine gunner. $15-$25 each.

Left to right: Breslin; three Japanese manufactured anti-aircraft guns, the third made of a rubber composition. L to R: $15-$20, $10-$12, $10-$12, $10-$12.

Anti-aircraft guns. Left to right: Molded Products (three versions); Playwood Plastics (two versions). $6-$10 each.

Barclay, observers with binoculars. Left to right: Two lying (long and short binocular versions); one kneeling; a Midi Series version. L to R: $15-$20, $45-$70, $20-$30, $45-$70.

Manoil observers. Left and right: two kneeling observers, standard issue (l) and 500 Series ®. Center: a lying observer with periscope and an aircraft spotter. L to R: $10-$15, $15-$25, $20-$25, $20-$25.

With binoculars. Left to right: Jones (three paint variants); Miller. L to R: $20-$22, $20-$22, $20-$22, $15-$25.

Kneeling with binoculars. Left to right: Grey Iron; Auburn Rubber (two paint variants); American Metal Toys (two versions, the second painted as enemy). L to R: $15-$20, $8-$12, $8-$12, $40-$60, $50-$100.

Kneeling with binoculars. Left to right: All Nu; Breslin; two Japanese manufacturers. L to R: $1000-$1500, $20-$25, $12-$15, $12-$15.

26. Anti-Tank

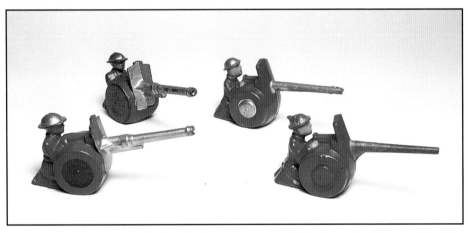

Anti-tank guns. Left to right: Barclay (two paint variants); American Metal (two versions). L to R: $10-$15, $10-$15, $40-$65, $40-$65.

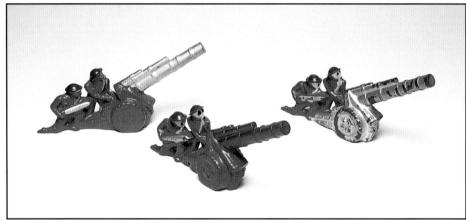

Barclay. 105 mm howitzer (three paint variants). $15-$20 each.

Playwood Plastic. Anti-tank guns (two versions). $6-$10 each.

Manoil. Left to right: round shield cannon; square shield version; cannon with wooden wheels. L to R: $20-$25, $20-$25, $30-$40.

Manoil. Man pulling trailer (metal and wooden wheel versions). L to R: $20-$30, $30-$35.

Manoil. Man pulling cannon (metal and wooden wheel versions). L to R: $20-$30, $30-$35.

Left to right: Barclay, kneeling shell loaders with clip-on helmet (l) and glued helmet (r); Manoil, standing figures, loading shell and fusing bomb. L to R: $10-$15, $10-$15, $10-$15, $20-$25.

Left to right: Barclay ammo carriers (two paint variants) and a prototype; Grey Iron, ammo carrier. L to R: $12-$17, $12-$17, prototype, $45-$60.

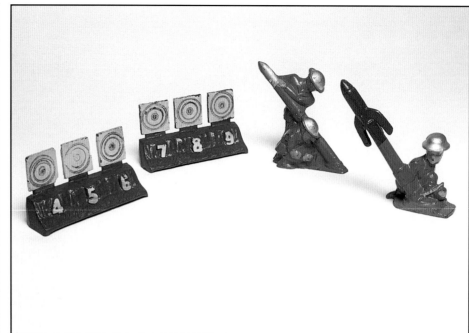

Left to right: Manoil, targets showing numbers four to nine; Barclay, mortar gun and team; Manoil, rocket launcher. L to R: $40-$65, $40-$65, $15-$25, $15-$20.

Ammo carriers. Left to right: American Metal Toys; Auburn Rubber; two Japanese manufactured hollow cast. L to R: $200-$300, $20-$30, $10-$15, $15-$20.

Auburn Rubber. Sound locator and mortar. $25-$40 each.

Left and right: Barclay, range finder (two paint variants). Center: Grey Iron, lying doughboy with range finder. L to R: $12-$18, $35-$45, $12-$18.

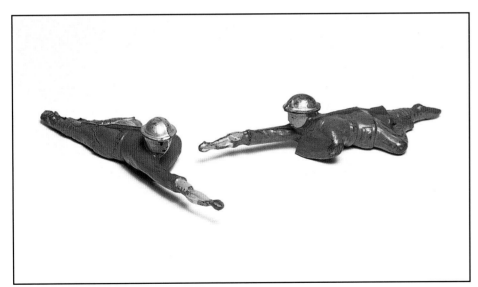

American Metal Toys. Prone wire cutters, second painted as enemy. L to R: $150-$250, $250-$300.

Manoil. Two versions of man carrying barbed wire. $25-$30 each.

Manoil. Paint variations of 45/46 Series mine detectors. $20-$25 each.

Digging soldiers. Left to right: Manoil; Barclay; Breslin. L to R: $30-$40, $25-$40, $25-$30.

27. Searchlights

Barclay. Three smooth lens versions and three ridged lens versions. L to R: $50-$75, $50-$75, $50-$75, $16-$24, $16-$24, $16-$24.

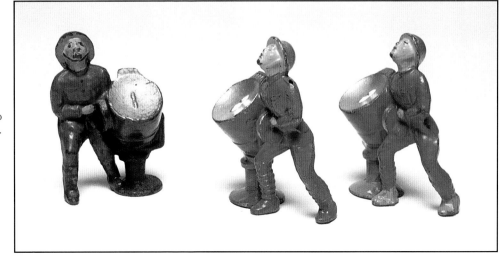

Left to right: Breslin; two Manoil versions. L to R: $15-$20, $20-$25, $15-$20.

American Metal Toys. Searchlights (two versions, the second has #27 on upright section). L to R: $55-$85, $85-$125.

Left to right: Barclay, searchlight and range finder; Auburn Rubber, searchlight; Barclay, sound detector. L to R: (Back) $20-$25, $18-$25. L to R: (Front) $20-$25, $20-$25.

28. Motorcycles

Left to right: Barclay, three versions; imitation of Barclay motorcycle with spoke wheels, (origin uncertain). $25-$40 each.

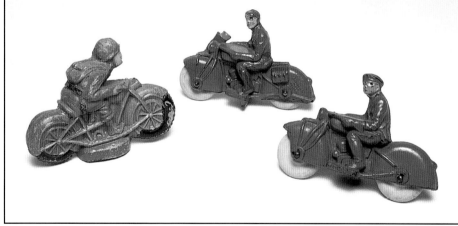

Left to right: Playwood Plastic; Auburn Rubber (two versions the first listed as Scout). L to R: $15-$20, $45-$60, $45-$60.

Manoil motorcycles. Back row (l-r): composition; standard issue. Front row (l-r): 500 Series; two standard versions. L to R: (Back) $30-$40 each. (Front) $35-$40, $30-$40, $30-$40.

Metal Cast. Motorcycle and rider in peak cap. $50-$75.

Motorcycle sidecar and machine gunners. Left to right: Barclay (two versions); Auburn Rubber. L to R: $50-$65, $50-$65, $45-$60.

Motorcycle machine gunner. Left to right: Breslin; American Metal Toys (two versions). L to R: $60-$75, $125-$150, $85-$125.

Manoil. Soldier riding cycle (two paint variants). $20-$30 each.

29. Accessories

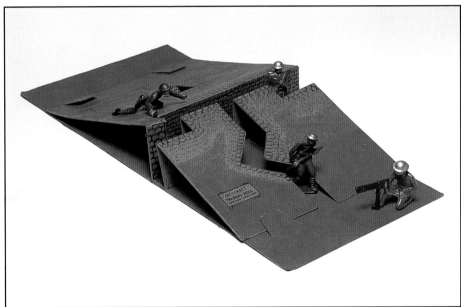

Above: Jaymar. Camp equipment in original box. $350-$450.

Top right: Grey Iron. Cardboard trench and soldiers. $150-$200 (trench only).

Bottom right: Left and right: Schneider Trench Company, composition trench sections. Center: Richard Appel, wooden victory tank. $60-$75, $50-$65 (tank).

Jaymar. Tents. $15-$45 each.

Sail-Me Company. Cardboard tents. $10-$15 each.

Sail-Me Company. Original box. $85-$100.

30. Ski Troops

Left to right: Barclay, khaki and white uniformed ski troops with automatic weapons; Grey Iron without tin skis; Barclay prototype; Manoil standing and lying ski troops. L to R: $50-$75, $35-$50, $50-$75, unique, $35-$55. (Front) $25-$40.

Grey Iron. Ski troop complete with tin skis. $50-$75.

31. Military School Cadets

Barclay. First and second series cadets. $10-$15 each.

Left to right: Barclay, short stride officer, cadet at slope, officer (paint variant), cadet painted as "toytown" figure, long stride (two paint variants); Manoil, hollow base and standard issue. L to R: $10-$15, $10-$15, $10-$15, $85-$100, $15-$20, $15-$20, $30-$50, $15-$20.

Left to right: Minikin, two versions in 1802 and 1952 uniforms; Jones (two paint variants); Lincoln Log. L to R: $15-$20, $25-$35, $15-$20, $15-$20, $12-$18.

Left to right: souvenir cadets, two versions in antimony; Beton; Barclay, first series drummer; unknown early manufacturer, possibly Soljertoy. L to R: $10-$15, $10-$15, $15-$20, $20-$30, $20-$25.

Opposite page:
Bottom right: Left to right: Grey Iron (two early paint variants); Port Arms, two slope arms and two West Point officers. L to R: $10-$12, $10-$12, $10-$15, $10-$15, $10-$15, $10-$15.

Left: Jones, three inch cadet. Right: Beton. L to R: $100-$125, $15-$20.

32. Air Force

Barclay. Left to right: Aviator; Mechanic with aircraft engine (two versions); Pod Foot (three paint variants); Pod Foot paint version of an Air Force officer. L to R: $10-$15, $25-$40, $25-$40, $25-$40, $7-$10, $7-$10, $7-$10, $50-$75.s

Manoil. Aviators (three paint variants). $15-$25 each.

Left to right: Manoil, aviators holding bombs (two versions, three paint variants); Manoil 500 Series (two paint variants); Breslin; Metal Cast. L to R: $15-$20, $15-$20, $15-$20, $20-$25, $20-$25, $15-$20, $65-$85.

Manoil. Left to right: aviators carrying aerial camera (two versions); three versions of men carrying propellers, the middle figure being the version with propeller away from head. L to R: $25-$30, $25-$30, $50-$75, $250-$375, $50-$75.

Aviators. Left to right: Grey Iron; Metal Cast; Jones; Japanese manufacturer; Molded Products (two versions). L to R: $20-$30, $40-$60, $18-$25, $20-$25, $8-$12, $8-$12.

Aviators. Left to right: Japanese manufacture, bisque; Jones copy in plastic; Grey Klip; Grey Iron, Uncle Sam's Defenders. L to R: $15-$20, $20-$30, $2-$3, $5-$7.

London Toy Canada. Two airmen and an aviator. $15-$20 each..

33. Parachutists

Barclay. Parachutists (three paint variants). $15-$20 each.

Parachutists. Left to right: Manoil (three paint variants); Breslin. L to R: $25-$30, $100-$150, $20-$25.

Manoil. Landing on cloud in helmet and landing on cloud in pilots hat. L to R: $50-$75, $55-$75.

Back row (l-r): Molded Products, four versions; an unknown aluminum example. Front row: two Playwood Plastic versions in foreground. Back L to R: $5-$8, $5-$8, $5-$8, $5-$8, $10-$12. Front $8-$12 each.

34. Navy

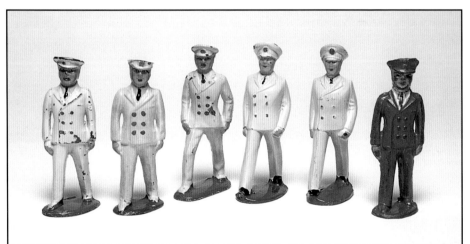

Naval officers. Left to right: Barclay, short stride, short stride tin top hat (missing), long stride; Manoil, Manoil hollow base, Japanese manufactured. L to R: $12-$18, $45-$70, $10-$15, $30-$50, $10-$15, $25-$35.

Left to right: Barclay, second version at the slope, two first versions at the slope and attention; Japanese manufactured, two sailors at slope. L to R: $8-$10, $10-$15, $10-$15, $12-$18, $12-$18.

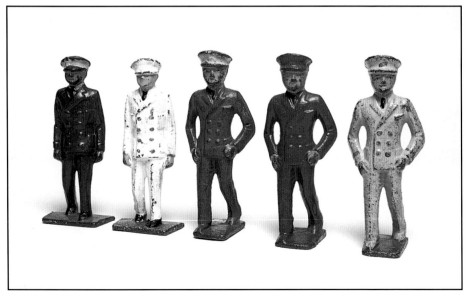

Naval officers. Grey Iron. Early versions in blue and white uniforms and three later versions. L to R: $8-$12, $8-$12, $10-$15, $10-$15, $10-$15.

Barclay. Left to right: two short stride in blue and white; long stride flag bearer; two bell bottom versions in white and blue; two sailors with puttees. L to R: $10-$15, $10-$15, $15-$25, $10-$15, $10-$15, $10-$15, $10-$15.

Grey Iron. Three early and two late versions in blue and white at the slope. L to R: $10-$12, $10-$12, $12-$15, $12-$15, $12-$15.

Left to right: Barclay, three post-war (paint variants), four Pod Foot in white and blue (paint variants); Manoil, hollow base in white; two Manoil standard issue paint variants. L to R: $15-$20, $15-$20, $15-$20, $7-$10, $7-$10, $30-$50, $10-$15, $10-$15.

Molded Products. Four versions at the slope. $5-$8 each.

Jones. Left to right: three-inch sailor at attention; 54 mm at attention (two paint variants); two paint and one casting variation at the slope; port arms ensign. L to R: $180-$200, $20-$22, $20-$22, $18-$20, $18-$20, $25-$30, $20-$25.

Japanese manufacture. Copies of American dimestores: naval officer, two ratings, and Marine officer. L to R: $30-$40, $20-$25, $20-$25, $35-$45.

Left to right: Jones, two at the slope, naval drummer, 1812 officer, seaman at attention; Lincoln Log sailor. L to R: $18-$20, $25-$30, $30-$35, $35-$50, $35-$50, $18-$22.

Degay. Sailors, possibly manufactured as souvenir items. $25-$35 each.

Manoil. Left: two paint variants of the Hot Papa Naval Firefighter. Right: two naval gunners (paint variants). L to R: $90-$135, $50-$70, $15-$20, $15-$20.

Barclay. Pirate with cutlass (two paint variants). $10-$15 each.

Jones. Left to right: Naval Academy; port arms; officer with binoculars; flag bearer; slope and port arm cadets. L to R: $20-$25, $20-$22, $35-$40, $20-$22, $20-$22.

Grey Iron pirates. Left to right: pirate with hook; pirate with dagger; pirate boy; pirate with two pistols; pirate with sword. L to R: $15-$20, $15-$20, $25-$30, $15-$20, $15-$20.

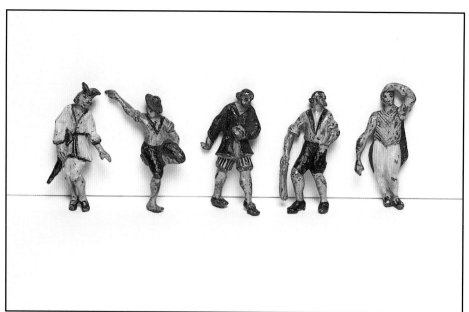

Unknown manufacturer. Cast iron pirates. $15-$20 each.

Divers with equipment. Unidentified manufacture, some copies of Manoil, others possibly by Japanese manufacturers. $15-$25 each.

Divers. Left to right: Barclay; Manoil (two paint variants); Manoil with no. 65 on chest. L to R: $350-$525, $15-$20, $15-$20, $15-$20.

Grey Iron. Left to right: lifeboatman at ship's wheel; comic fisherman with pipe and fish; fisherman with net. $25-$35 each.

35. Marines

Barclay. Left to right: second series saluting, port arms, officer and slope; two first series officer and slope. L to R: $12—$18, $7-$10, $10-$12, $10-$12, $8-$10, $10-$12.

Barclay. Left to right: short stride; long stride; cast helmet Marines. L to R: $30-$40, $10-$15, $45-$65.

Marines marching at the slope. Barclay. Left to right: two short stride versions, first with tin top hat; long stride; long stride post-war; Pod Foot. L to R: $50-$75, $15-$20, $15-$20, $20-$30, $7-$10.

Left to right: Molded Products, four examples, first and second version blue uniforms, and first and second version white uniforms; H.B. Toys; unidentified aluminum; Japanese manufacturer, hollow cast Marine. L to R: $5-$8, $5-$8, $7-$12, $7-$12, $12-$18, $10-$12, $12-$15.

Left to right: Manoil, at the slope; Grey Iron, first and second version port arms; Auburn Rubber, port arms with early and late version Marine buglers. L to R: $10-$15, $20-$25, $10-$15, $12-$18, $10-$15, $10-$15.

Jones. Left to right: three-inch figure port arms; Marine Calvary; port arms; officer with binoculars; slope (two paint variants); 1812 Marine. L to R: $180-$200, $75-$100, $20-$22, $20-$22, $20-$22, $20-$22, $18-$20.

36. Foreign Troops

Barclay. Left to right: Italian infantryman and officer; Japanese infantryman and officer; Chinese officer and infantryman. L to R: $100-$150, $75-$140, $75-$90, $110-$165, $120-$170, $80-$120.

Left to right: Barclay, Ethiopian tribesman and barefoot officer; Grey Iron, rifleman, tribesman with sword, two Ethiopian infantrymen and officer. L to R: $100-$150, $125-$190, $30-$45, $30-$45, $30-$45, $30-$45, $30-$45.

American Metal Toys, enemy troops. Back row (l-r): Slope arms; grenade thrower; falling wounded; Front row (l-r): officer pointing; advancing with gas mask; standing firing; advancing with rifle. L to R: (Back) $75-$100, $90-$150, $225-$300. (Front) $150-$200, $75-$100, $200-$250, $175-$250.

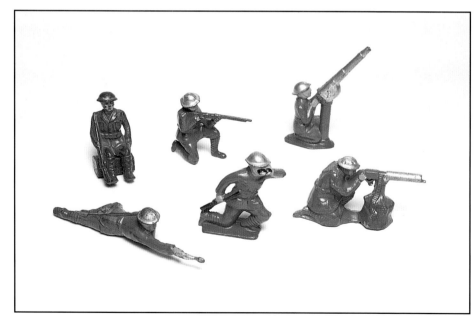

American Metal Toys, enemy troops. Back row (l-r): seated; kneeling firing; anti-aircraft gunner. Front row (l-r): wire cutter; man with binoculars; seated machine gunner. L to R: (Back) $60-$80, $50-$100, $60-$80. L to R: (Front) $250-$300, $50-$100, $60-$80.

Left to right: Jones, World War I German at the slope, officer with binoculars, standing firing; American Metal Toys/Jones lying firing. L to R: $200-$225, $200-$225, $175-$200, $80-$125.

American Metal Toys/Jones. World War I German (the two figures on the right in slightly lighter colored uniforms may possibly have been meant to represent Finnish troops). Charging $100-$125. Kneeling $80-$100.

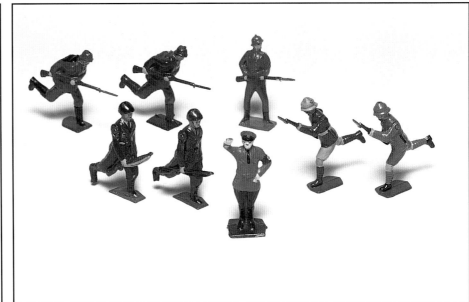

Jones. 54 mm charging and standing Germans, running Italians, Adolf Hitler and two British Colonial charging troops. L to R: (Back) $25-$30, $25-$30, $25-$30. (Middle) $20-$25, $20-$25, $25-$30, $25-$30. (Front) $35-$40.

Left to right: Barclay, mounted New York Militia; Highlander; Jones, British foot guard; Auburn Rubber "Palace Guard" officer and three guards (paint variants). L to R: $25-$35, $10-$15, $20-$25, $12-$18, $12-$18, $12-$18, $12-$18.

Grey Iron. British infantryman and officer. L to R: $95-$140, $85-$130.

Grey Iron. Greek Evzone. $50-$75.

Jones. Greek Evzone (four paint variants). L to R: $25-$30, $25-$30, $25-$30, $35-$45.

Left to right: Barclay, Royal Canadian Mounted Policeman with pistol; two Lincoln Log Royal Canadian Mounted Police. L to R: $10-$15, $55-$65, $15-$18.

Grey Iron. Late and early versions port arms and mounted Royal Canadian Mounted Police. L to R: $20-$25, $12-$18, $25-$40, $20-$35.

Above: Left to right: Barclay, first series Legion officer; Grey Iron, machine gunner, advancing and slope arm Legion; Paul Paragine Legion figure at the slope. L to R: $10-$12, $20-$25, $20-$25, $20-$25, $50-$60.
Top left: Souvenir. Royal Canadian Mounted Policemen. $20-$30 each.
Below: Minikin. Left to right: U.S. Horse Marine 1902; French Indochina; French garrison gunner; French Senegalese trumpeter; French ski trooper; French Algerian with standard; French Chasseur with standard; French Artillery. L to R: $30-$35, $20-$25, $20-$25, $20-$25, $25-$30, $20-$25, $20-$25, $20-$25.

37. Miscellaneous

Japanese manufacturer. Copies of American-made Dimestore soldiers. $35-$65 each.

Barclay. Pod Foot examples in red: grenade, bazooka, firing rifle, and crawling with pistol. $40-$75 each.

Metal Cast. Flag bearer and officer. L to R: $60-$85, $50-$70.

Warren. Although probably never sold through the five and ten cent stores, Warren figures are highly collectible. Cavalry figures. $100-$130 each.

Warren. Infantry with plug in heads and moveable arm. $60-$80 each.

Part III

Historical and Non-Military

38. Historical Figures

Top left: Taller figures, left to right: American Metal Toys, knight with shield; Barclay, knight with shield and knight with lance; American Metal Toys, knight with lance. Shorter figures: Barclay, Pod Foot knights. L to R (taller): $75-$100, $10-$15, $10-$15, $75-$100; (shorter): $15-$20 each.

Bottom left: Left to right: Minikin, mounted knights (two paint variants); John Hill and Company, England, Crusader; Jones, Archer. L to R: $15-$20, $35-$40, $25-$30, $15-$20.

Top right: Minikin. 14th – 15th century knights. L to R: $15-$20, $20-$25, $15-$20.

Minikin. Norman knights (six paint variants) with original box. $25-$30 each.

Left to right: Japanese manufacturer, plaster composition knights; Grey Iron, knight; John Hill and Company, England, Crusader. L to R: $8-$10, $10-$15, $8-$10, $10-$15, $35-$40.

Japanese manufacturer, Sonsco. Knights in original box. $40-$50.

Minikin. Hannibal's War Elephant set. $250-$300 set.

Minikin. Left to right: Samurai warrior; Miamotono Yoritomo mounted; Kato Kiyomn Asa. L to R: $25-$30, $60-$80, $25-$30.

Minikin. Left to right: Archer of the Nitano Shiro Tribe; Anayama Kosuko; Kinoshita Tokichiro; Tamerlane; Benkei, the fighting monk; Warrior of the Kinoshita Tribe. $25-$30 each.

Minikin. Napoleon and Henry the 4th. $75-$80 each.

Grey Iron. Colonial flag bearer. $150-$200.

Left to right: Grey Iron, Colonial soldier and officer; Lincoln Log, Abe Lincoln with axe, Fort Ticonderoga and Fort Dearborn soldiers. L to R: $15-$20, $15-$20, $20-$25, $20-$25, $20-$25.

Left to right: Lincoln Log, George Washington mounted (two versions); Grey Iron, George Washington mounted (two paint variants); Metal Cast, George Washington mounted. L to R (Back) $30-$35 each. L to R (Front) $30-$40, $30-$40, $35-$45.

Historical Miniatures. Colonial soldier and flagbearer. L to R: $60-$80, $50-$70.

Minikin. The Spirit of '76. $40-$65 set.

185

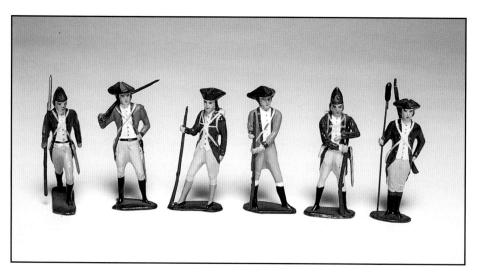

Minikin. Left to right: American Revolution; British 40th Foot Regiment; Green Mountain Ranger; Pennsylvania Regiment; Green Mountain Ranger private; Haslet's Delaware Regiment; Captain John Lamb's artillery. L to R: $20-$30 each.

Jones. Left to right: Wayne's Legion (two paint variants); Hessian (two paint variants); unidentified regiment. L to R: $20-$25, $20-$25, $15-$18, $15-$18, $30-$35.

Jones. Left to right: British Grenadiers 1775 (three paint variants); British grenadiers with ramrod (two paint variants). L to R: $20-$25, $20-$25, $20-$25, $18-$22, $18-$22.

Jones. American War of Independence officers (seven paint variants). $20-$25 each.

Jones. Various colonial regiments. $30-$35 each.

Jones. French 1763 regiments (four paint variants). $30-$35 each.

Jones. Left to right: American Marines at ease (four paint variants); Colonial regiments (two paint variants). $18-$22 each.

Jones castings. Left to right: four British, American War of Independence; American 1812 artillery man; rifleman, 1781. $15-$25 each.

Jones. Left to right: American War of Independence French infantryman and British artillery man; American infantry 1812 (two paint variants); 1836; American War of Independence British infantry. L to R: $25-$30, $30-$35, $20-$25, $20-$25, $25-$30, $25-$30.

Jones, three-inch scale. Left to right: British Infantry Drummer; Hessian Grenadier; American Marine Officer. L to R: $180-$220, $150-$175, $180-$220.

Jones, three-inch scale. Left to right: British Marine at trail and slope; British Light Infantryman; British Grenadier. L to R: $150-$175, $150-$175, $150-$175, $150-$175.

Jones, three-inch scale. Left to right: British Marine firing; French Foreign Legion on guard; Highland Piper; Highlander charging. L to R: $200-$225, $150-$175, $175-$200, $175-$200.

Jones. Three inch scale American Marine and six inch scale American Marine. L to R: $20-$25, $350-$450.

Jones. British Marines 1812 (three paint variants). $20-$25 each.

Left to right: Wilton, souvenir of Fort Ticonderoga; Metal Cast, American Revolution, French, British, and American troops. L to R: $20-$25, $30-$35, $30-$35, $30-$35.

Jones. Left to right: charging highlanders, 1781 (three paint variants); charging Highland officer. L to R: $18-$20, $18-$20, $18-$20, $30-$35.

Jones. Left to right: Highlander, 1814; Highlander, 1776 with fixed bayonet; two Highlanders, slung rifle; Highlander WWI uniform; two Highland pipers in kilt and trews. L to R: $18-$20, $30-$35, $18-$20, $18-$20, $25-$30, $30-$35, $30-$35.

Minikin. Left to right: Gordon Highlanders Officer, 1914; 74th Highlander Drummer, 1914; Kings Own Scottish Borderers Officer, 1686; 74th Regiment Officer, 1846; Black Watch piper, 1815. $30-$35 each.

Kresge. Composition British infantry man and officer, probably made in Germany and supplied to Kresge's five-and-ten cent stores. L to R: $25-$30, $50-$75.

Minikin. British Dragoon, 1850, and British Coldstream Guard, 1742. L to R: $40-$50, $20-$25.

Jones. American Civil War. Union and Confederate troops advancing and at attention. L to R: $20-$22, $20-$22, $18-$20, $18-$20, $18-$20.

Jones. Zouaves charging (five paint variants). $20-$25 each.

Wilton. Left to right: Confederate officer; Union and Confederate soldiers, fixed bayonet. $15-$20 each.

Home Cast. Paul Revere. $20-$25.

Jones. Colonial man and woman, paint variants. $40-$45 each.

Top right: Jones. Colonial man and woman, possibly intended to represent The Belle of New York, the Belle of Baltimore, the Dandy of Charleston, or The Dandy of Philadelphia. $40-$45 each.

Center right: Historical Miniatures, lead. Left to right: George Washington, Martha Washington, Abraham Lincoln, F.D. Roosevelt, Winston Churchill, and Chiang Kai-Shek. $30-$35 each.

Bottom right: Historical Miniatures, composition. Left to right: MacArthur, General Montgomery, Franco, Patton, and DeGalle. $30-$35 each.

39. Circus

Grey Iron. Grey Craft Clever Clowns Circus with net in original box. $350-$500. Single items, $15-$20 each.

Top right: Nifty, Japanese manufacturer. Circus figures, original box lid.

Bottom right: Nifty circus figures in box. $1200-$1500.

Top left: Japanese manufacture. P.T. Barnum (third right), Tom Thumb (bottom left), and other circus figures influenced by Britains circus line. L to R: $35-$50, $35-$50, others $25-$40.
Top right: Left to right: Auburn Rubber, circus animals and clown (suffering from some rubber fatigue); Metal Cast clown (second right). Animals $30-$35, Clown $35-$45, Metal Cast $25-$30.
Bottom: Barclay Circus wagon and horse-drawn circus animal cage. L to R: $125-$150, $200-$250.

40. Disney

Lincoln Log. Three of the Seven Dwarfs: Doc, Bashful and Grumpy. $65-$85.

Japanese manufacturer. Snow White and the Seven Dwarfs. *Courtesy Debbie Bednarek*. $150-$175.

Home Cast. Reproductions from original Home Cast molds. Minnie Mouse, Pluto and Mickey Mouse, Three Little Pigs, Baby Mickey Mouse and long-billed Donald Duck. *Courtesy of Eccles Brothers Ltd.* From original molds by Home Foundry. Current retail price.

Japanese manufacturer. Bambi and Butterfly. $65-$75.

41. Character Figures

Top left: Tommy Toy, Nursery Rhyme. Left to right: Little Miss Muffet; Humpty Dumpty; Puss in Boots; Old King Cole; RARE Jack and Jill; Little Bo Peep; Jack and the Beanstalk; Old Mother Hubbard; Tom Tom the Piper's Son; Old Mother Witch. All $60-$75, except Jack and Jill, $1000-$1200, and Old Mother Witch, $100-$140.

Top right: Tommy Toy, Nursery Rhyme. Left to right: Old Mother Hubbard; Puss in Boots; Tom Tom the Piper's Son; Jack and the Beanstalk; Humpty Dumpty; Little Bo Peep (paint variants) $60-$75 each.

Bottom right: Tommy Toy, Nursery Rhyme. Left to right: Little Miss Muffet; Puss in Boots; Old Mother Witch (paint variants). L to R: $60-$75, $60-$75, $140-$180.

Unknown manufacturer. Cast iron Boy Blue. $50-$60.

All Nu. Dogs' band in red suits. $50-$60.

All Nu. Dogs' band in white suits. $50-$60 each.

Unknown manufacturer, possibly Tommy Toy or All Nu. Cop and hooker salt and pepper shakers. $60-$80 pair.

Left to right: Manoil, Toy Town drummer prototype; Grey Iron, The Royal Guard, also available in blue; Barclay Toy Town soldier based on the West Point Cadet figure. L to R: unique, $75-$100, $85-$100.

Jones. Robinson Crusoe. $50-$60; casting $25-$35.

Lincoln Log, Og Son of Fire, issued as premiums for Libby's milk. Left to right: Three Horn; Nada; Og; Ru; Rex; Big Tooth. L to R: $50-$65, $50-$65, $50-$65, $50-$65, $75-$100, $50-$65.

Home Cast. Left to right: Dick Tracy; Junior; Chief Brandon; Boris Arson; Pat Patton; Tess Truehart. *Courtesy Eccles Brothers Ltd.* From original molds by Allied Manufacturing. Current retail price.

Home Cast. Left to right: Bluto; Popeye; Olive Oil; Little Orphan Annie; Sandy; Daddy Warbucks; The Gumps; Andy; Hilda; Uncle Bim. *Courtesy Eccles Brothers Ltd.* From original molds by Allied Manufacturing. Current retail price.

Home Cast. Left to right: Moon Mullins, Kayo, Uncle Willy, Harold Teen, Shadow, Pop Jenks, Smitty, Herbie and The Boss. *Courtesy Eccles Brothers Ltd.* Current retail price.

42. Space and Science Fiction

Britains. Buck Rogers series commissioned by and manufactured on license by Buck Rogers creators J. Dille. Left to right: Wilma; Ardala; Buck; Killer Kane; Dr. Huer; Mekkano Robot. $250-$350 each.

Tootsietoy. Buck Rogers space craft. $200-$250 each. Boxed example $300-$350.

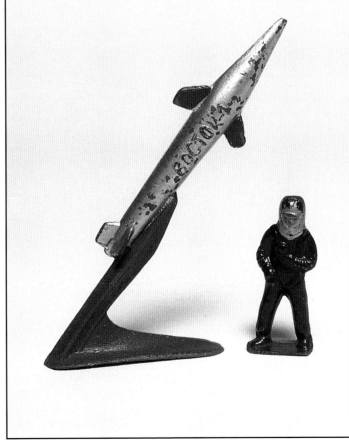

Left: unknown manufacturer, rocket and launcher; right: Japanese manufactured spaceman. L: $40-$50; R: $25-$30.

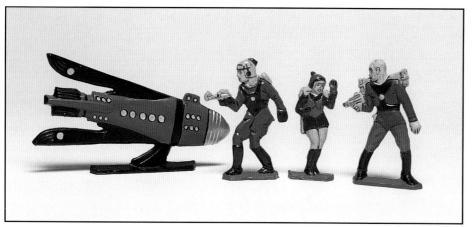

Home cast. Left to right: Buck Rogers space ship; Buck Rogers; Wilma Deering; Killer Kane. *Courtesy Eccles Brothers Ltd.* From original home cast molds by Junior Caster. Current retail price.

Home cast Buck Rogers series. Amphibious Neptunians and Neptunian Sub-Ship. *Courtesy Eccles Brothers Ltd.* From original home cast molds by Junior Caster. Current retail price.

Home cast Buck Rogers series. Left to right: Depth Man; Buck Rogers; Dr. Huer; Black Barney; Disintegrator; Tieko Man. *Courtesy Eccles Brothers Ltd.* From original home cast molds by Junior Caster. Current retail price.

Home cast Buck Rogers series. Left to right: Mekkano; Interplanetary Spaceship; One Eyed Man; Tiger Man; Tiger Ship; Asteride. *Courtesy Eccles Brothers Ltd.* From original home cast molds by Junior Caster. Current retail price.

Home cast Flash Gordon series. Left to right: Dale Ardan; Flash Gordon; Dr. Zarkov; Ming the Merciless; Hawkman; Prince Barin. *Courtesy of Eccles Brothers Ltd.* From original home cast molds by Home Foundry. Current retail price.

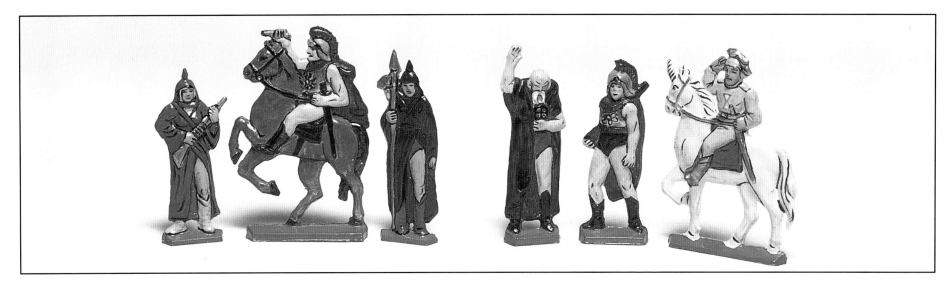

43. Souvenir and Novelty

Japanese manufacturer. Souvenirs from Fort Ticonderoga, Fort William Henry and Fort Niagara. $15-$20 each.

Left to right: Japanese manufacturer, General Jackson riding Sam Patch, a souvenir from The Hermitage, Tennessee; General Lee, a souvenir of Gettysburg. L to R: $30-$35, $25-$30.

Left to right: Japanese manufacturer, Fort Niagara souvenir; Wilton; Confederate souvenir of Horseshoe Curve; Japanese manufacturer, WWI doughboy souvenir and WWII private; souvenir of Camp Wolters, Texas; highlander souvenir from Windsor, Canada. L to R: $15-$20, $15-$20, $10-$15, $10-$15, $20-$25.

Japanese manufacturer. Pierot couple, a souvenir of Washington, D.C. $40-$50 pair.

Japanese manufacture. Souvenir Indians war dancing. $15-$20 each.

U.S. and Japanese manufacturers. Various cowboy souvenir items. $15-$35 each.

Japanese manufacture. Souvenir mounted Indians. $15-$25 each.

Japanese manufacturer. Souvenir Indians. $15-$35 each.

Japanese manufacture. Souvenir desk accessories: paperweights, pen and ink holders. $20-$30 each.

Japanese manufacturer. Souvenir Indians. $15-$35.

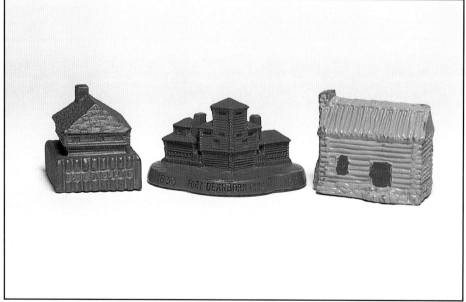

U.S. or Japanese manufacturers. Fort Souvenirs. The log cabin is hollow cast of unknown origin and is probably not a souvenir item. L to R: $35-$45, $35-$45, $20-$25.

Multi Products. George Washington. Six inch high wooden folk art style souvenir. $50-$65.

Multi Products. Soldier, sailor and airman. Composition. $50-$60 each.

Manoil. Mr. Prohibition prototype, possibly designed as a bottle stopper. *Courtesy Eccles Brothers Ltd.* Unique.

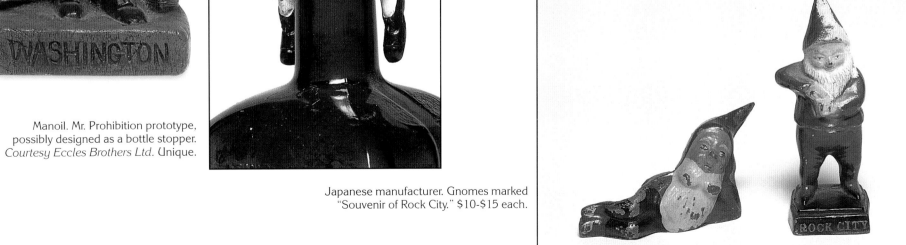

Japanese manufacturer. Gnomes marked "Souvenir of Rock City." $10-$15 each.

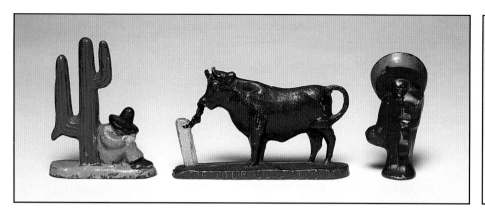

Unknown manufacturer. Left to right: cast iron Mexican sleeping under cactus; bull tied to post; hollow cast Mexican. $20-$25 each.

All Nu. Baby skunk and Elmer Skunk with (scent)/cent. L to R: $15-$20, $50-$60.

Japanese manufacture. Pack mules with prospector. These pack mules surface all over the U.S.A. at various national monuments and tourist attractions. $5-$10 each.

U.S. or Japanese manufacturers. Souvenir pack mules and prospectors panning for gold. $10-$15 each.

All Nu/Faben. Souvenir racehorses and a cowboy horse. $15-$20 each.

Top: All Nu/Faben, souvenir elephants. Bottom (l-r): unknown manufacturers, cast iron tiger; lion; bear cub. Elephants, $50-$65; others, $10-$15.

44. Sports

Barclay baseball players. Left to right: batter; pitcher; catcher. $75-$95 each.

Unknown manufacturer, solid cast baseball players. Left to right: fielder; pitcher; catcher. $15-$20 each.

Unknown manufacturers. Left to right: pitcher; cast iron umpire; Japanese bisque baseball batter. L to R: $60-$75, $35-$40, $15-$20.

Auburn Rubber baseball players with red caps. Left to right: fielder; pitcher; catcher; batter. L to R: $15-$20, $20-$25, $20-$25, $20-$25.

Auburn Rubber baseball players with blue caps. Left to right: catcher; base runner (two versions); batter. L to R: $20-$25, $15-$20, $15-$20, $20-$25.

Unknown manufacturer. Baseball players in aluminum, possibly modern manufacture. $5-$10 each.

Japanese manufacturer. Little League baseball player set tied into original box. Imported into the U.S.A. by Shackman NY. $100-$125 set.

Sheila football players. Back: backs. Front (l-r): lineman; center; linemen (two paint variants). $65-$75 each.

Auburn Rubber football players. Left to right: carriers; passers; linemen; backfieldsmen; centers (two paint variants). (Left top) $25-$30 each. (Left front) $15-$20 each. (Middle) $20-$25. (Right top) $25-$30. (Right front) $20-$25.

Sheila football players. Left to right: solid, hands on hips and quarterback passing; hollow cast quarterback passing; solid runningback. L to R: $65-$75, $65-$75, $85-$100, $65-$75.

Left to right: unknown manufacturer, composition runningback; Hubley, kicker; unknown manufacturer nickel-plated runningback. L to R: $15-$20, $25-$35, $20-$25.

Comet Authenticast. Football players in paint variants. Eleven piece set: $100-$125 (unboxed).

Back (l-r): home cast, runningback, tackler, kicker; Japanese manufacturer, runningback (two paint variants). Front (l-r): unknown manufacturer, two centers from a board game. L to R: (back) $6-$8, $6-$8, $6-$8, $10-$15, $10-$15; (front) $10-$15 each.

Comet Authenticast. Football game. Two teams. Eleven piece set: $100-$125 each (unboxed).

Unknown manufacturer. Football players from board game. $4-$6 each.

All Nu/Faben. Racehorses and jockeys in different colored silks and numbered saddle cloths. The figure in the foreground is listed as "Hunter on horse" and has a hunting horn. $25-$35 each.

Barclay. Jockeys and horses, paint variants with different numbered saddle cloths. $15-

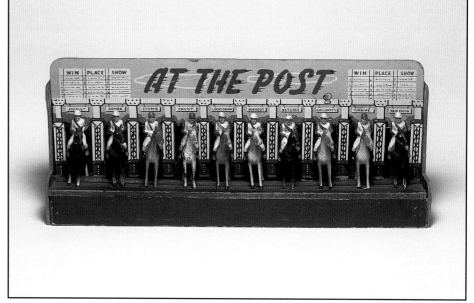

All Nu. Post-war racehorse display with wooden starting gate, "At the Post." The racehorses contained within this set are those that featured in the Suffolk Turf Handicap run at Suffolk Park, Boston, Massachusetts in 1946: (1) New Mood, (2) Turbine, (3) Gallorette, (4) Natchez, (5) Mahlout, (6) Lucky Draw, (7) Pavot, (8) Stymie, (9) Armed, (10) Assault. *Courtesy Marv Breitlow.* Unique.

All Nu. Male and female jumpers. Paint variants. $30-$35 each.

Unknown American and Japanese manufacturers. Racehorses and jockeys, probably made for the souvenir market. $10-$20 each.

Back (l-r): All Nu, jockey and racehorse, and larger scale racehorse; Japanese manufacturer racehorse and jockey. Front (l-r): jockey on galloping horse, unknown manufacturer and possibly by Soljertoy. L to R: (back) $45-$50, $25-$30, $20-$25; (front) $25-$35, $20-$25.

Unknown American and Japanese manufacturers. Racehorses and jockeys souvenirs. $10-$20 each.

Unknown American and Japanese manufacturers. Racehorses and jockeys issued as souvenirs. $10-$20 each.

Japanese manufacturers. Sulky racers (four paint variants) and miniature version. $10-$12 each.

All Nu. Sulky racers (three paint variants). $30-$40 each.

Left to right: All Nu, polo player; unknown American manufacturer, possibly Soljertoy, three polo players (paint variants). L to R: $30-$35, $35-$40, $35-$40, $35-$40.

Japanese manufacturer. Left to right: two ten pin bowling; athletic sprinters with original box; boxer. $10-$12 each unboxed, $15-$20 boxed.

Uncertain origin, possibly by All Nu or Breslin. Left to right: ice hockey player; ten pin bowler; football quarterback; football running back. *Photograph courtesy of Don Houde.* (valuation not yet established).

45. Army Vehicles

Manoil (left to right). Back row: tractor; caisson; rocket launcher with wooden wheels; caisson with metal wheels. Row two: camp stove; water wagon (three versions). Row three: gas canister (two paint variants); gasoline trailer. Front row: torpedo; bomb; pontoon boat. (Back row) $20-$25 each. (Second row) $20-$25 each. (Third row) $25-$30 each. (Front row) L to R: $30-$35, $18-$20, $35-$45.

Back row (l-r): Tommy Toy, cannon truck; Barclay, metal wheel cannon truck; Tommy Toy cannon truck (paint variant); Barclay armored car (post-war version). Middle row (l-r): Tommy Toy, cannon truck with working headlight and battery holder; Barclay, anti-aircraft car (post-war version); Barclay armored car (pre-war version); Barclay armored car (post-war paint variant). Front row (l-r): Barclay, small cannon truck (pre- and post-war versions). Barclay armored gun car (pre-war version). L to R: (Back row) $30-$35, $25-$30, $30-$35, $20-$25. (Middle row) $100-$125, $20-$22, $20-$25, $20-$25. (Front row) $20-$25 each.

Back row (l-r): Barclay, cannon trucks (early and late version); Manoil, cannon trucks (two paint variants). Middle row (l-r): Barclay cannon truck; Breslin (copy of Manoil); Barclay cannon truck (late pre war version). Front row (l-r): Manoil, gun truck; Barclay two-man machine gun cars (two paint variants). L to R: (Back Row) $40-$50, $35-$45, $35-$45, $35-$45. (Middle row) L to R: $35-$40, $35-$50, $65-$75. (Front row) L to R: $30-$35, $45-$55, $45-$55.

Back row (l-r): Manoil, siren truck (two-piece casting version); Barclay, searchlight trucks (two paint variants); Manoil, siren truck (one piece casting version). Middle row (l-r): Barclay, anti-aircraft trucks (two paint variants). Front row: American Metal Toys pill box bunkers (four paint variants). L to R: (Back row) $65-$75, $80-$90, $80-$90, $55-$65. (Middle row) L to R: $20-$25, $20-$25. (Front row) L to R: $75-$80, $75-$85, $85-$100, $75-$85.

Back row (l-r): Barclay, cannon on wheels (pre war version), U.S. Army truck (pre-war versions), cannon on wheels (post war version), U.S. Army truck (post-war version). Second row (l-r): All Nu, searchlight, sound detector and cook stove on wheels; Barclay, tractor (pre-war version). Third row (l-r): Barclay, cannon (pre-war olive drab version), motor units (post- and pre-war versions), cook stove on wheels (pre war version). Front row (l-r): Ralstoy, cannon and armored car; Barclay cook stove on wheels and U.S. Army truck (olive drab versions). L to R (Back) $18-$20, $15-$20, $20-$22, $20-$22. (Second row) R: $30-$35, $30-$35, $30-$35, $20-$25. (Third row: $18-$20, $18-$20, $18-$20, $20-$22. (Front: $18-$20, $25-$30, $20-$22, $20-$22.

Back row (l-r): American Metal Toys, anti-aircraft tank; Manoil tank; Barclay small tank (post war). Second row (l-r): American Metal Toys, tank with flame thrower; Manoil tank (paint variant); Ralstoy tank with wooden hubs for rubber tread. Third row (l-r): Barclay small tank (pre war); American Metal Toys flame thrower tank (paint variant); Manoil tank (paint variant); unknown manufacturer, possibly Metal Cast, tank. Fourth row (l-r) Barclay small tank (pre-war paint variant); American Metal Toys flame thrower tank (casting version); All Nu tank-like vehicle. Front row (l-r): Japanese manufactured tank; Arcade tank with rubber tracks. L to R (Back row) $90-$110, $20-$25, $18-$20. (Second row) $100-$125, $20-$25, $30-$35. (Third row) $20-$25, $100-$125, $20-$25, $30-$35. (Fourth row) $20-$25, $100-$125, $30-$35. (Front row) $25-$30, $60-$80, $30-$35.

Back row (l-r): Unknown manufacturer, composition tank; Auburn Rubber one man tanks (late and early paint variants). Second row (l-r): Barclay Renault tank; Barclay (post-war, two paint variants); Well Made Doll Company composition tank. Front row (l-r): Metal Cast Tank; Arcade tank and cannon; Auburn Rubber, small one man tank. L to R: (Back row) $25-$30, $30-$40, $30-$40. (Middle row) $25-$35, $20-$25, $20-$25, $15-$20. (Front row) $35-$45, $150-$175 pair, $20-$25.

Back row (l-r): Victory wooden tank and original box. General Grant tank and original box. Middle row (l-r): unknown manufacturer wooden tank; Sun Rubber tank with moveable turret; unknown manufacturer wooden tank. Front row (l-r): Barclay "4562" Tank (one and two man versions paint variants). L to R: (Back row) $50-$60, $65-$75. (Middle row) $15-$20, $75-$100, $10-$15. (Front row) $35-$45 each.

Top left: Back: Chein tin Army truck.. Middle row (l-r): Auburn Rubber army truck (two paint variants); Barr Rubber Ford army truck. Foreground: Ralstoy transport truck with aircraft. (Back) $125-$150. (Middle row) $35-$45 each. (Foreground) $65-$75.

Top right: Left: Well Made Doll Company, Jeep with machine gun. Back right: Sun Rubber large armored car. Front right: London Toy, Canada, flat bed army truck. Left: $20-25; back, $75-$90; front, $30-$35.

Bottom right: Back: Well Made Doll Company large Jeep. Foreground (l-r): Ralstoy cannon and searchlight truck; Barclay motorized howitzer (pre-war); unknown manufacturer wooden cannon truck. (Back) $35-$50. L to R: $75-$85, $20-$25, $10-$15.

Top left: Back row: Barr Rubber, Ford ambulances (three paint variants). Middle row: Barclay ambulance (large version); Sun Rubber ambulances (two paint variants). Front row: Barclay small ambulances (two painted cross variants); Sun Rubber ambulance (paint variant). (Back) $45-$55 each. (Middle) L to R: $75-$85, $30-$35, $30-$35. (Front) L to R: $35-$45, $35-$45, $30-$35.

Top right: Back row (l-r): London Toy, Canada, metal wheel cannon; Barclay cannon with elevation handle; London Toy, Canada (wooden wheel version). Middle row (l-r): Manoil metal wheel cannon, howitzer, and wooden wheel cannon. Front row (l-r) Manoil metal wheel cannon (casting variant); Hahn and Grey Iron cannons. (Back row) L to R: $25-$30, $50-$60, $25-$30. (Middle row) L to R: $10-$15, $10-$12, $10-$15. (Front row) L to R: $10-$15, $20-$25, $25-$30.

Bottom right: Back row: Auburn Rubber, howitzer; American Metal Toy cannon. Middle row (l-r): Barclay cannon (large version); London Toy, Canada, cannon; Barclay cannon (medium version). Front row (l-r): unknown manufacturer wooden cannon; Barclay cannon (small version). L to R: (Back row): $35-$40, $50-$60. (Middle row): $35-$40, $25-$30, $20-$30. (Front row) $10-$15, $20-$25.

Back row (l-r): American Metal Toy anti-aircraft gun; Japanese manufacturer tin cannon; Barclay swivel cannon with spring firing mechanism. (Foreground) Barclay coastal gun and searchlight. (Back) L to R: $35-$45, $25-$30, $40-$50. (Foreground) L to R: $65-$85, $40-$50.

Back: Manoil, Action Cannons with original boxes. Foreground (l-r): Manoil Action Cannons (two paint variants); Breslin copy of Manoil without firing mechanism. (Back Row) $35-$45 each with box. (Foreground) $20-$25, $20-$25, $30-$35.

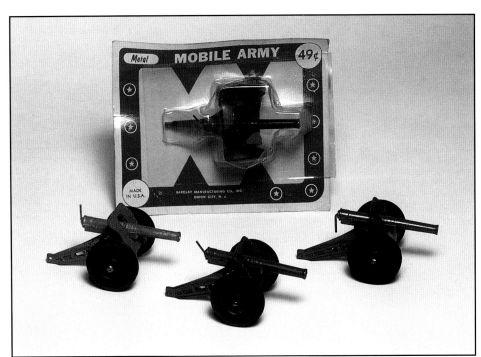

Barclay. Post-war cannons. Original blister pack and three paint variants. (Back) $85-$100. (Front) $25-$30 each.

Grey Iron. Cap firing machine gun. $150-$300.

46. Aircraft

Back row (l-r): Sun Rubber, army bomber; Arcor Toy/Auburn Rubber Jet 559 (post-war). Second row (l-r): Auburn Rubber Flying Boat; Sun Rubber army bomber; Auburn Rubber fighter plane. Third row (l-r): Auburn Rubber "TWA" commercial airliner and four engine transport plane; unknown manufacturer, possibly Sun Rubber dive bomber. Fourth row (l-r): Auburn Rubber fighter plane (paint variant) and four engine transport plane (paint variant). L to R: (Back row) L to R: $35-$40, $50-$60. (Second row) $60-$75, $35-$40, $30-$35. (Third row) $40-$45, $40-$45, $30-$35. (front row) $30-$35, $45-$50.

Back row: Sun Rubber, "Pursuit Ship" and Civilian (paint variant). Second row: Sun Rubber "Pursuit Ship" (three paint variants). Third row (l-r): Barclay, "U.S." army plane; London Toy, Canada, Hurricane Hawker; Barclay "U.S." army plane (paint variant). Front row (l-r): Ralstoy "U.S. Army" plane; Barclay piggy back airplane; Barclay U.S. Army bomber with original bombs and clip. (Back Row) $30-$35 each. (Second Row) $30-$35 each. (Third Row) L to R: $35-$45, $30-$35, $35-$45. (Front Row) $30-$35, $75-$80, $80-$100.

Barclay. Airship. $125-$150.

47. Ships

Well Made Doll Company. Left to right: aircraft carriers (two sizes), destroyers (two sizes) made out of composition material due to WWII metal storage. L to R: $40-$50, $25-$35, $35-$40, $25-$35.

Left to right: Well Made Doll Company, submarine, torpedo boats (small and large versions); Velasco Toy Company, "Minneapolis" weapon submarine.. (Back) L to R: $35-$40, $30-$45, $40-$50, $25-$30

Opposite page:
Back row (l-r): Manoil submarine; Barclay battleship; Ralstoy battleship. Second row (l-r): unknown manufacturer, possibly Barclay, "USS New Mexico" battleship; Auburn Rubber battleship. Third row (l-r): Auburn Rubber submarine; Japanese manufacturer "Texas" battleship souvenir item; Manoil submarine (paint variant). Foreground (l-r): Unknown manufacturer aircraft carrier; Barclay aircraft carrier with two original aircraft. L to R: (Back Row) $30-$35, $35-$45, $35-$45. (Second Row) $50-$60, $35-$45. (Third Row) $30-$40, $20-$25, $30-$35. (Foreground) $20-$25, $75-$85.

48. Box Art Work, Catalogs, Advertising, Molds

Barclay. Original mold for unproduced sentry. *Courtesy Eccles Brothers Ltd.* Unique.

Barclay and Manoil molds. *Courtesy Eccles Brothers, Ltd.* Unique.

Top left: Barclay. Box artwork and soldier set. $600-$750.

Top right: Barclay. Cellophane fronted boxes containing Pod Foot Cowboy and Indian sets. $275-$300 each.

Bottom right: Barclay. Midi series blister pack and cellophane fronted box containing Pod Foot set. Midi Series: $500-$550. Pod Foot: $300-$350.

Grey Craft. Grey Iron packaging with clips holding pieces in place on backing card. $300-$350.

Auburn Rubber. Box art. $15-$20.

World Soldiers Series box art work. $15-$20.

Molded Miniatures. Jones box art. $15-$20.

Tootsie Toy. American Infantry boxed set. *Courtesy Kent Wood.* $65-$85.

Japanese manufacturer. Best Maid soldier set and box art. $75-$85.

Jones. Paint Them Yourself British Guardsman complete with paints and brushes. $35-$45

True Craft Models. Original box and solid cast figures. $8-$10 each.

Barclay. Paint Your Own Army Set with original box. $200-$250.

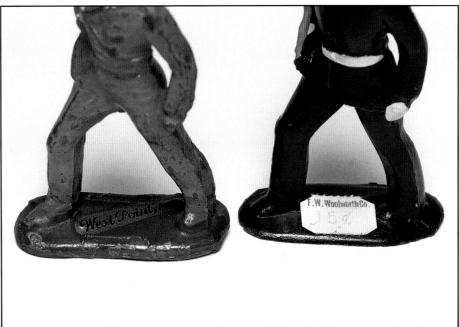
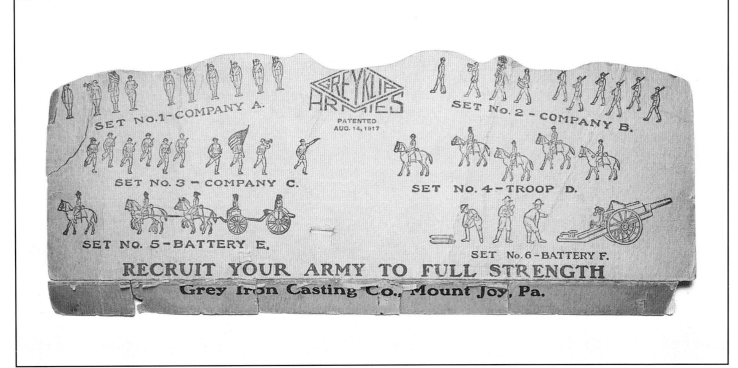

Top left: Grey Klip Armies, Grey Iron, backing card and figures. $150-$200.

Top right: Barclay. Original labels, the first showing decal attached to base of West Point Cadet figure sold as a souvenir at West Point Academy and an original F.W. Woolworth 5 cent sticker attached to the base of a Barclay Marine. $20-$25 each.

Bottom left: Grey Klip Armies, Grey Iron, reverse of backing card. $150-$200.

Factory tags. Jones molded miniatures and Barclay. Unique.

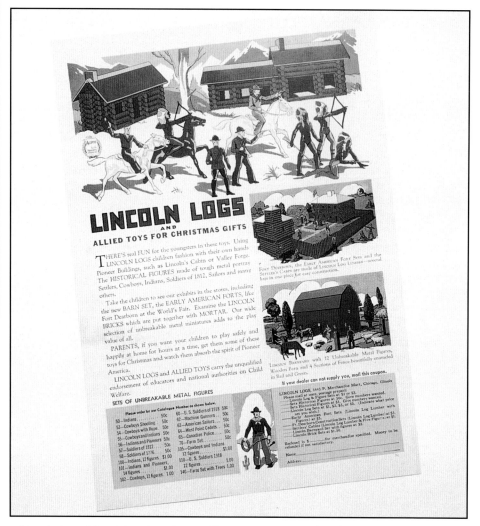

Lincoln Log. Advertising sheet from a Christmas magazine. $25-$35.

Jones. Colonial Lady Figures with original paper label DeLuxe figurine. $50-$60.

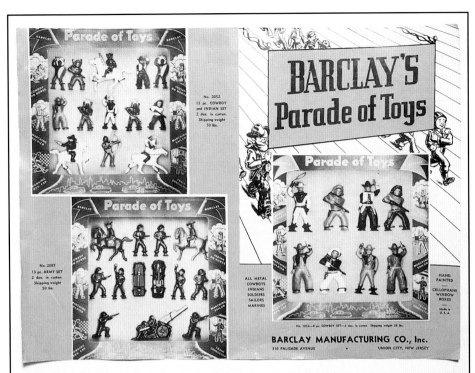

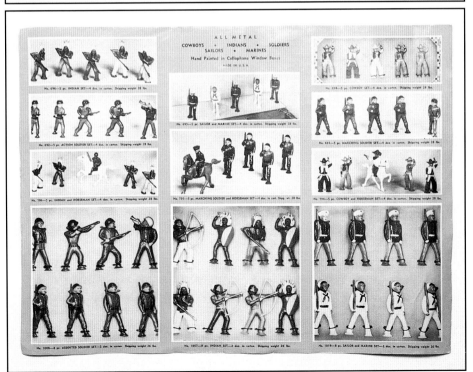

Top left: Barclay. Parade of Toys catalog front cover. $25-$35.

Bottom left: Barclay. Parade of Toys catalog paper. $25-$35.

Top right: Manoil. Catalog page. $25-$35.

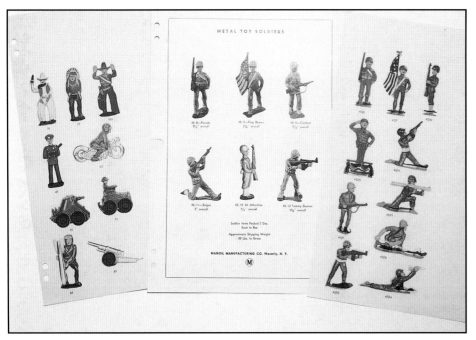

Manoil. Color catalog pages. $25-$35.

Home Foundry. Catalog. $20-$30.

Manoil. My Ranch black and white and color catalog pages. $25-$35.

Manoil. Original illustrated labels from factory bulk pack boxes showing Happy Farm series. $2 each.

Manoil. Original illustrated labels from factory bulk pack boxes showing Happy Farm series. $2 each.

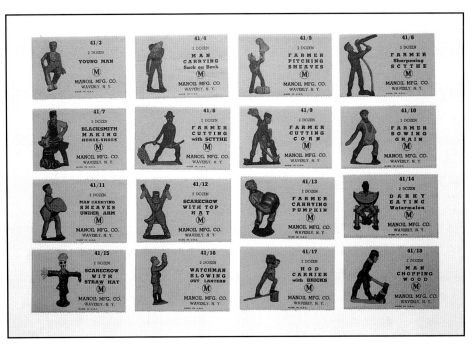

Manoil. Original illustrated labels from factory bulk packs showing Military series. $2 each.

Manoil. Original illustrated labels from factory bulk pack boxes showing Happy Farm series. $2 each.

SUBJECT INDEX

Advertising Items, 236, 238, 239
African American Family, 22, 23
Aircraft, 226, 227
Aircraft Carriers, 228-229
Airmen, 162-163
Airship, 227
Ambulances, 223
American Civil War, *see Historical*
American Family, 20-25
American Legion, 70
American War of Independence, *see Historical*
Amish Family, 23-24
Animals
  Bears, 209
  Bulls, 37, 63
  Calves, 40, 41
  Cat, 42
  Cows, 37, 38, 40, 41
  Deer, 43
  Dogs, 35, 38, 43
  Donkeys, 38, 40
  Elephants, 209, *see also Circus*
  Foals, 39
  Goats, 42
  Horses, 37, 38, 39, 40, 209
  Lambs, 37, 41
  Lion, 209
  Pack Mules, 40, 208
  Pig, 37, 38, 42
  Piglets, 37, 42
  Poultry, 38, 42, 43
  Sheep, 37, 38, 41
  Skunk, 208
  Tiger, 209
  Turkeys, 42, 77
Armored Cars, 218, 220
Artillery, *see Soldiers and Vehicles*
Art Work, 231-236
Athletics, 217
Aunt Holly, 73
Aviators, *see Airmen*
Bambi, 196
Bandits, *see Cowboys*
Backwoodsmen, 62
Baseball, 210-212
Beach, *see American Family*
Belle of Baltimore, 192, 236
Belle of New York, 192, 236
Black Cats, *see Halloween*
Bluto, 200
Bowling, 217
Boxing, 217, *see also Soldiers*
Boy Blue, 198
Boy Scouts, 69, 113
Boxed Sets, 20, 26, 27, 33, 36, 37, 38, 50, 51, 61, 62, 64, 157, 158, 183, 193, 200, 212, 217, 221, 225, 231-235
Brakeman, 26
Braves, *see Indians*
British Troops, 175, 190, *see also Historical*

Bucking Broncos, 59, 205
Buck Rogers, 201-203
Burgler, 69
Cadets, 160-161
Camp Equipment, 84, 157-158
Camp Wolters, 204
Canada, *see Royal Canadian Mounted Police*
Cannons, 218-220, 223-225
Carpenters, 35
Catalogs, 237-239
Cheng Ki Shek, 192
Chiefs, *see Indians*
Children, 21-23, 26, 28-30
China-Indo, 177
Chinese Troops, 173
Christmas, 71-73
Churchill, Winston, 192
Circus, 193-194
Circus Animals, 193-194
Circus Wagons, 193, 194
Clowns, 193, 194
Coastal Gun, 224
Colonial Troops, *see Historical*
Comic Characters, 200-201, *see also under Characters Name*
Conductors, 26, 27
Confederate Troops, *see Historical*
Cop and Hooker, 198
Covered Wagons, 62
Cowboys, 53-63, 205
Cowgirls, 60
Crusoe, Robinson, 199
Dandy of Charleston, 192, 236
Dandy of Philadelphia, 192, 236
DeGaulle, Charles, 192
Detective, 69
Dick Tracey, 199
Disney, 195, 196
Divers, 170
Dogs, *see Animals*
Dogs Band, 198
Donald Duck, 196
Dwarfs, 195
Easter Rabbit, 77
Elmer the Skunk, 208
Enemy Troops, *see Foreign Troops and Soldiers*
Engineers, 26-28
Ethiopian Troops, 174
Evzones, 176
Factory Samples and Ephemera, 238-239
Farm, 33-43
Farmers, 34, 36-38
Farmers Wives, 36, 37, 42
Farm Workers, 33-35
Father Christmas, *see Santa Claus*
Father Time, 11
Fences, 20, 37, 63
Firemen, 67

Fishermen, 170
Flag Bearers, *see Historical and Soldiers*
Flash Gordon, 203
Football (American), 212-214, 217
Foreign Troops, *see under Country Name*
Fort Dearborn, 206
Fort Niagara, 204
Fort Ticonderoga, 204
Fort William Henry, 204
Forts, 206
Garageman, 21, 66
German Troops, 174, 175
Girls Band, 65
Gnomes, 207
Greek Troops, 176
Gumps, The, 200
Guns, 224-225
Gunslingers, *see Cowboys*
Halloween, 76
Hannibal, 183
Happy Farm, 33-35
Highlanders, *see Foreign Troops and Historical*
Historical
  American Civil War, 191
  American War of Independence, 184-189, 191
  Colonial Troops, 185-188
  Confederate Troops, 191
  Highlanders, 188-190
  Knights, 182, 183
  Personalities, 185-192, 207
  Union Troops, 191
Hitler, Adolf, 175
Hobos, 39
Horse Racing, 214-216
Hostess, 65
Hot Papa, *see Navy*
Houses, 206
Humpty Dumpty, 197
Ice Hockey, 217
Indo China, *see China*
Indians, 44-52, 204-206
Italian Troops, 173, 175
Jack and Jill, 197
Jack and the Beanstalk, 197
Jackson–General and Sam Patch, 204
Japanese Troops, 173
Jockeys, 214-216
King Henry 4th, 184
Knights, 182-183
Labels, 239
Lee, General, 204
Legion, 177, *see American Legion*
Lifeboatmen, 170
Lincoln, Abraham, 185
Little Bo Peep, 197
Little Miss Muffet, 197
Little Orphan Annie, 200

Lone Ranger, 62
MacArthur, General, 94, 192
Mailmen, 66
Majorettes, 66
Marines, *see also Historical*, 171, 172, 187-189
Medieval, *see Historical*
Mexicans, 208
Mickey Mouse, 196
Military, *see Soldiers and Vehicles*
Military School Cadets, *see Cadets*
Military Vehicles, *see under Vehicle Name*
Milkman, 21
Minnie Mouse, 196
Molds, 35, 230
Montgomery – Field Marshall, 192
Moon Mullins, 200
Motor Cycles
  Military, 155, 156
  Police, 58
Mr. Prohibition, 207
Musicians, *see Soldiers*
My Ranch, *see Cowboys*
Napoleon, 184
Nativity, 71
Naval Academy, *see Navy*
Navy, 18, 166-169
Newsboys, 21, 29
New York Militia, 15, *see also Foreign Troops*
Nurses, *see Soldiers*
OG – Son of Fire, 199
Oilers, 27
Old Father Time, 11
Old King Cole, 197
Old Mother Hubbard, 197
Old Mother Witch, 76, 197, *see also Halloween*
Olive Oil, 200
Pack Mules, *see Animals*
Paint Your Own Soldier Sets, 234
Parachutists, 164-165
Passengers – Railroad, 26, 28-31, *see Railroad*
Patton, General, 192
Pill boxes, 219
Pirates, 169, 170
Pluto, 196
Policemen, 67-69
Police Motorcycles, 68-69
Polo Players, 216
Popeye, 200
Porters, 26-27, 31-32, *see Railroad*
Pumpkins, 76
Puss 'n Boots, 197
Ranch, 22, 63
Revere, Paul, 191
Rocket Ships, *see Space*
Rodeo, *see Cowboys and Souvenirs*

Roosevelt, F.D.R., 192
Royal Canadian Mounted Police, 176-177
Sailors, *see Navy*
Samurai Warriors, 184
Santa Claus, 71, 72
Scarecrows, 33, 34
Science Fiction, 201-203
Scottish Troops, *see Foreign Troops and Historical*
Ships, 228, 229
Shoeshine Boy, 29
Skaters, 74, 75
Skiers, 74, 75
Skunk, *see Animals and Elmer the Skunk*
Sleds, 74
Sleighs, 72, 73
Snow White, 195
Soldiers Advancing, 17, 134-138, 178
  Airborne Infantry, *see also Parachutists*
  American Cavalry Circa 1900, 14, 15, 17, *see also Cavalry*
  American Civil War, *see Historical*
  American Infantry Circa 1900, 14, 16, 17, 18
  American Legion, 70
  American War of Independence, *see Historical*
  Ammo Carriers, 150
  Anti-Aircraft Gunners and Guns, 144-146
  Anti-Tank Gunners and Guns, 148, 149
  Artillery, 142-149, *see also Vehicles*
  At ease, 101
  Attacking, 138-139, 142-143
  Attention, 101, 102
  At the Ready, 14, 16, 101, 102
  Automatic weapons, 138
  Barbed Wire Carriers, 151
  Bazookas, 140-141, 178
  Binoculars, 146, 147, *see also Observers*
  Boat – Soldiers in, 119
  Boxers, 83, *see also Sport*
  Buglers, 104-105, 107-108
  Bullet Loaders, 131
  Cameramen, 83
  Camp Scene, 84, 157
  Cavalry, 14, 15, 110, 114-117, 179
  Charging, 132-139
  Chefs, 80, 84
  Clarinets, 107-108
  Colonial, *see Historical*
  Communications, 118-119

  Crawling, 137
  Cyclists, 155
  Cymbals, 108
  Digging, 152
  Doctors, 87, 88
  Dogs – Soldiers with, 103
  Drum Majors, 104, 107, 108, 178
  Drummers, 14, 16, 17, 105-108, 178
  Eating, 81, 84
  Enemy Troops, 81, 89, 90, 93, 97, 120, 123, 130, 132, 136, 145, 151, *see also Foreign Troops*
  Field Telephone, 118
  Fifes, 17, 107, 108
  Firing
    Kneeling, 18, 121-123, 178
    Lying, 124, 178
    Standing, 120-122, 125, 178
  Flagbearers, 14, 109-111, 179, 184, 185, *see also Signal Flags, Cavalry, Navy and Colonial*
  Flamethrowers, 140
  Flare Guns, 143
  French horn, 107
  Gas masks, 142-143
  Grenade Throwers, 132-133, 143, 178
  Howitzers, 145, *see also Artillery and Vehicles*
  Lewis Gun, 146
  MacArthur, 94, *see also Historical*
  Machine Gunners, 126-131, 143
  Marching, 17, 18, 95-100, 178
  Medics, 87
  Mine Detectors, 152
  Mortar, 150
  Motorcycles, 155, 156
  Musicians, *see Individual Instruments*
  Navy, 18, 166-169, *see also Signal Flags*
  New York Militia, 14, 15, 175
  Nurses, 85, 86, 88, 91
  Officers, 16, 92-94, 178, 179
  Observers, 146-147, *see also Binoculars*
  Paymaster, 82
  Photographers, 83
  Pigeon Handler, 119
  Port Arms, 100-101
  Present Arms, 101
  Radio Operators, 118-119
  Range Finders, 151
  Recoil Rifle, 131
  Rocket Launchers, 150, 201
  Sailors, *see Navy, Signal Flags and Flag Bearers*

  Searchlights, 153, 154, *see also Vehicles*
  Sentries, 103
  Sentries with Dogs, 103
  Shell Loaders, 149
  Signal Flags, 112-113, *see also Boy Scouts and Navy*
  Siren truck, 219
  Sitting, 81-82, 84
  Ski Troops, 159
  Sound Locators, 148, 154
  Sousaphone, 107, 108
  Stretcher Bearers, 89-91, 143
  Stretcher Parties, 89-91, 143
  Surgeons, 87, 88
  Targets, 150
  Tin Plate Soldiers, 19
  Tommy guns, 137-138
  Trombone, 108
  Tubas, 108
  Typing, 80, 81, 84
  Walkie Talkies, 119
  West Point Cadets, *see Military School Cadets*
  Wire Cutters, 151
  Wounded, 87-91
  Writing, 82, 84
Souvenirs, 204-209
Space, 201-203
Spaceships, 201-203
Spirit of '76, 185
Sport, *see under Individual Sports*
St. Patrick's Day, 77
Sulky Racing, 216
Tanks, 220-221
Tee Pee, 50
Tents, 84, 157-158
Thanksgiving, 77
Three Little Pigs, 196
Tinplate Soldiers, 19
Tonto, 62
Tom Tom the Pipers Son, 197
Toy Town Soldiers, 199
Trees, 37
Trench Sections, 157
Trotter Racing, *see Sulky Racers*
Uncle Mistletoe, 73
Uniforms – Non-military, 65-70
Union Troops, 191
Vehicles – Military, 218-225, *see under Vehicle Name*
Washington, George, 185, 192, 207
Washington, Martha, 192
Wedding Figures, 24, 25
Western Union, 66
Wig Wam, 50
Winter Figures, 73-75
Witches, 76, 197
Zouaves, *see Historical*